A Home
From the Woods

A Home From the Woods

◆

Adventures and methods
restoring and building
authentic log cabins

by
Michael Antoniak

with illustrations
by Lucas Antoniak

Writers Club Press
San Jose New York Lincoln Shanghai

A Home
From the Woods
Adventures and methods
restoring and building
authentic log cabins

All Rights Reserved © 2002 by Michael Antoniak

No part of this book may be reproduced or transmitted in any form or by any means, graphic, electronic, or mechanical, including photocopying, recording, taping, or by any information storage retrieval system, without the permission in writing from the publisher.

Writers Club Press
an imprint of iUniverse, Inc.

For information address:
iUniverse, Inc.
5220 S. 16th St., Suite 200
Lincoln, NE 68512
www.iuniverse.com

ISBN: 0-595-24571-4

Printed in the United States of America

This book is dedicated to the memory of
John Williamson Park
who first inspired Liz and I
to dream of our own log cabin

and to his grandchildren
Marcus, Lucas, Sarah, Sean, Matthew and Peter
the heirs of our dreams

Contents

Introduction xiii

PART I: Restoring An Authentic Log Cabin

| CHAPTER 1 | A Log Cabin Home of Our Own .... 3 |
| CHAPTER 2 | Methods for Uncovering and Restoring An Old Log Cabin .... 23 |

- *Finding Your Cabin* 23
- *Tools* 26
- *Uncovering The Cabin* 27
- *Cleaning The Walls* 29
- *Repairing/Replacing Logs* 32
- *Treating The Logs* 34
- *Preparing To Chink* 36
- *Applying The Chinking* 38
- *Modernizing The Cabin* 39

PART II: The Hand-hewn Log Cabin

| CHAPTER 1 | A Hand-Hewn Addition, The Old Fashioned Way .... 43 |
| CHAPTER 2 | Methods for Building A Hand Hewn Cabin .... 65 |

- *Selecting The Site* 66
- *The Logs* 68

vii

- *Selecting Your Trees* *69*
- *When To Cut?* *71*
- *Bringing Them Down* *72*
- *Cutting The Log* *74*
- *Moving The Logs* *75*
- *Hewing The Logs* *76*
- *Hewing With An Ax* *78*
- *Working With An Adze* *79*
- *Notching The Corners* *81*
- *Make A Template* *82*
- *Planning For Electrical And Plumbing* *86*
- *Starting The Cabin* *89*
- *Framing The Door* *90*
- *Raising The Logs* *90*
- *Framing The Windows* *93*
- *The Cap Logs* *94*
- *The Roof* *95*
- *Final Steps* *97*

Part III: Building A Cabin With Round Logs

| **Chapter 1** | A "Quick and Easy" Cabin .... 103 |
| **Chapter 2** | Methods for A Rounded Log Cabin .... 115 |

- *Planning* *116*
- *Cutting Your Logs* *117*
- *Prepping The Site* *118*
- *The Rounded Saddle Notch* *119*
- *The Rest Of The Cabin* *121*
- *The Chinking* *123*

Part IV: Removing and Rebuilding An Original Log Cabin

Chapter 1	Trials and Tribulations With An Unexpected Find	127
Chapter 2	Methods for Taking Down and Rebuilding A Log Cabin	159

- *Finding A Salvageable Cabin* *159*
- *Document Before Dismantling* *159*
- *How Will You Use It?* *160*
- *Take Plenty Of Pictures* *160*
- *Keep Copious Notes* *161*
- *Assess The Structure* *162*
- *Isolate The Cabin* *163*
- *Removing The Chimney* *163*
- *Tagging The Logs* *164*
- *Preparing To Take Down The Cabin* *167*
- *Final Preparation* *168*
- *Undoing The Roof* *169*
- *Taking Down The Log Walls* *170*
- *Lowering The Logs* *171*
- *Windows And Door Frames* *171*
- *Removing Floors, Joists And Sills* *172*
- *Putting It All Back Together* *173*
- *Replacing Damaged Logs* *174*
- *Adding A Loft for Upper Rooms* *176*
- *Windows, Doors, Interior Walls* *176*
- *Framing The Roof* *177*
- *Modifying Or Adapting An Old Cabin For New Uses* *177*
- *Cleaning The Logs* *179*
- *Treating The Logs* *180*
- *Chinking* *181*

- *Finishing Touches, Final Thoughts*................................ *181*

Epilogue.. 183

APPENDIX A   Tools 187

APPENDIX B   Adapting The New: Modern Methods and Materials In A Log Cabin 201

APPENDIX C   Other Methods for An Authentic Looking Log Cabin 211

Resources 215

Acknowledgment

Many people helped in some capacity, directly or indirectly, by act or encouragement, on our log cabin projects. For that we're ever grateful.

Some are gone, some have moved on, and some remain a part of our lives.

For fear of leaving anyone out, I'll simply thank you all here.

Introduction

Let me begin with absolute candor. Growing up in New York City, log cabins didn't mean a thing to me.

And if you asked me what I knew of Tennessee the morning Liz and I were to be married, all I could recall was the specter of "Volunteers" swinging from the drapes at the White House as they partied after the inauguration of Andrew Jackson. That's something that stuck with me from grade school history.

My father-in-law drove all night from Nashville to our wedding. He was one of what Southerners refer to as a "damn Yankees"—a New Yorker who moved South and stayed. In the brief time we had to speak on our wedding day, he advised us to consider relocating to Tennessee. He was sure we would like the state and the climate, could find work, and might welcome the change in scenery.

Over the next year he wrote Liz with some regularity, always pushing this Southern migration. Then, the following spring, 1977, John and his wife Joyce bought a 10-acre "farm" in the country with an old dilapidated house on it. He had been told it was a log cabin, but wasn't sure. He wrote to see if we'd be interested in trying to fix the old place up, as they were planning to build a home on another part of the property.

Liz and I had talked often about living in the country, and had been getting the old Strout Realty catalog of rural property for several months, pondering the possibilities. This seemed like a dream come true, and once he sent us pictures, we started planning our move.

What we could see of the house through the weeds held little more than potential, but the view from it's porch was simply unbelievable.

We left New York for Tennessee in late June, and that's really how we got started on log cabins.

Over the past 25 years Liz and I have worked together on four different log cabin projects, all the while raising six children, assorted goats, dogs, cats, chickens with a few pigs and cows thrown in.

We've learned a lot, never stop learning, about country living and everything related to it. Down through the years we've mastered all sorts of skills we never envisioned, and gained an intimate understanding of the cycles of nature.

But this book's about our cabin experiences. Each of its four main sections covers a different type of log cabin project. The first half of each is a straightforward narrative telling our story, and experiences. In the second part of each section I explain the methods we used, and you can adopt or adapt for a similar project.

So you *really* want to want to build a log cabin?

Read on!

You'll find the inspiration and know-how, or a reality check —whichever you need— somewhere in these pages.

PART I

Restoring An Authentic Log Cabin

1

A Log Cabin Home of Our Own

To us, it was an old house, thoroughly neglected, with potential. The farm had been on the market for years, until the real estate agent duped some outsiders who knew so little about the value of local property they would eagerly pay more than its worth.

That was my father-in-law John and his wife Joyce. They saw something in the property none of the locals recognized. The view directly opposite the house—a succession of intersecting hills framing a valley stretching to the west—is as good as you'll find anywhere. But they weren't concerned about the condition of the old house anyway. Their plans were to retire to a secluded house they would build on the hill across the road.

John thought about tearing the old place down, but then offered to sell it and a couple of acres to us.

"It's supposed to be a log cabin," he wrote us in New York, and enclosed a picture of the house and view. We knew so little then we mistook the clapboard siding for logs.

That was in May 1977. Six weeks later, on the hottest day in July, we were standing in the yard, for first impressions of what we would make our home. Everywhere we looked were the symptoms of abandonment: broken windowpanes, a rotting porch, weeds towering over us, trees growing right up against the back door.

I remember walking around the entire house, then pausing in the yard to get a feel for the place. It had very good vibes. Liz and I smiled at each other. "We can do something with it," we agreed. John laughed.

As soon as we cut down a cedar tree grown up against the back door, we made our way inside. What we found there made the outside look well cared for: broken remains of furniture, the stuffing torn out; heaps of old clothes; stacks of rotting newspapers—whatever someone could leave or drag there scattered in every room. The roof over one half the house had rotted through, and years of rain had rotted the floor beneath it.

Despite the filth, we figured we could salvage the living room, even though its ceiling was barely 6 feet high; the room directly behind it; a smaller room to the right; and the "kitchen" on the left, roughly 8x8 with no sink and only one electrical outlet.

"How can you tell it's a log cabin?," I asked but John shrugged.

"That's what I've been told," was still the best he could offer. We poked around the outside, and thought we could see something that looked like a log behind a short piece of loose clapboard hanging by the front window.

A week later Liz and I officially moved to the country when John dropped us off with our camping gear. We slept in a 5x7 tent the first few nights, afraid to sleep in the house for all the rat traffic we could hear, even during the day. That first week we managed to clean the place out, emptying the house of junk and piling it in the yard. We scrubbed the floors and walls with water we hauled in backpacks, by the gallon, from the spring in the woods. After a few days we felt the place was clean enough to move in. We cooked by camp stove and lantern. It was tough falling asleep at night for the sounds of rats racing back and forth in the walls and ceiling. I slept with a machete within reach but never had the occasion to use it.

We didn't own a car yet, and the real estate agent recommended a "hippie" who might haul off the junk which filled the yard for a few bucks. A few days later "Tom", a drop-out from New Jersey, showed up in a car filled with kids. He gave lip service to talk about helping with any work, and was more interested in touring the old house.

"Got yourself a log cabin," he smiled, standing by the fireplace in the living room.

"How can you tell?" we wanted to know. He explained that people in these parts used to be ashamed of their log cabins, so they would cover them up at first chance they could. "Even paint the rock fireplace a shiny silver, for that modern look," he laughed, pointing at the edges of the stones which showed behind the mantle.

"But there's a couple of ways to know if it's a cabin," he continued. "First look at the walls. The logs they used to build cabins were at least six inches thick. By the time you cover them up you've got a wall that can be ten inches, or a foot thick." He led us over to the window and door and showed what he meant by the thickness of the frames.

The other way to tell, he continued, was by the shape of the room. The typical cabin was 16x18 around here, he explained, and the ceilings tend to be low in the old cabins. The old timers usually added a loft sometime after the cabin was built to make extra space, usually for sleeping. That often left the ceiling in the downstairs room as low as they could stand it.

"Got a crowbar," he asked. "I'll show you."

We watched as he forced dents into the plasterboard walls, working the mantle loose from the wall it hid. "Sure enough, there's your logs," he smiled, pulling the mantle forward enough to reveal the ends of several logs. They were tacked to a wooden frame that outlined the hearth and stone of the chimney. We were absolutely thrilled.

He walked around the room a moment and looked up around the chimney, then picked up the crowbar again. "There has to be a loft upstairs," he assured us. "Want to take a look?"

You bet we did! Tom moved to the front corner nearest the mantle and window and knocked a hole in the drywall ceiling. "The stairs to the loft were usually in a corner," he explained. Soon as the bar up hit more wood ceiling he moved to the adjacent corner. This time he struck his first hole along the outside of a rafter and contents of a rat's nest spilled to the floor.

"One time I was tearing out the ceiling in an old house, and a pair of black snakes fell out of the ceiling and onto me," he shared. "I've tried to be a little more cautious ever since."

He moved closer to the corner and knocked another hole. This time dust drifted to the floor. He poked the crowbar all the way up.

"Here's the loft," he grinned, and proceeded to tear enough of the drywall from the ceiling, exposing a hole so we could see into the attic. He called one of the boys he'd brought with him, and asked if he wanted an adventure. Before the kid could accept or decline Tom pushed him up through the hole.

"What do you see?," he wanted to know. Then, not satisfied with the answer, or confident there were no snakes, Tom pulled himself up through the hole.

"I bet no one has been up here in years," he exclaimed, and we were the next to join him. We helped Liz into the loft, then I scrambled my way up the wall and joined them.

"See the logs, there's two courses above the floor," he said, We could make out the two layers of square logs in the light slipping in through gaps in the clapboard. As our eyes adjusted to the light we could make out the shapes of chairs, piles of books, magazines, the odds and ends you'd expect to see in an attic. Beads of light slipped in through nail holes in the tin roof, helping reveal the small trees and rough logs which served as the rafters for the roof.

After a while we all climbed back down for another look at the logs before pushing the mantle back in place. Tom promised to come by next week to haul off the junk but I don't think we saw him again for three or four years. By then, he'd found Jesus and was driving around with another crop of kids in a banged up van. He had crudely spray painted "Jesus is Love" in red on both sides and across the front.

That day, though, after he left, we went to work tearing down the rest of the ceiling. Above the drywall we found rough sawn ceiling rafters with uneven boards nailed to them to make the loft floor. Between the second and third rafters, where scatterings of a rats nest had spilled out, we uncovered a rats nest that ran the full 16 foot length of the room. It was crammed with whatever rats thought worth collecting during the many years since that drywall ceiling had been built: tatters of quilts, twigs, buttons and buckets of pecan, hickory and walnut hulls.

When the weather cooled we found we were losing too much heat through the ceiling gaps between the old boards so we put a layer of particle board down in the attic. That was pretty much all the progress we made restoring our cabin until the following spring, but we passed a lot of time those cold months dreaming of uncovering of our cabin.

Getting there took us through winter, which came on quicker than we'd anticipated. We were so "green" as our neighbor Doug continually reminded us, before winter set in it never occurred to us the house didn't have a thermostat or central heat. The limited electricity—one outlet and light switch in every room—which we had turned on a month after moving in, and the fact we had to haul our water in gallon jugs, didn't bother us. But the lack of heat, eventually got out attention.

As a gift John bought us an old wood stove and showed us how to set it up. We were grateful, as we never let him know how limited our means were. We spent a good part of that winter with a 17-inch bow saw struggling to feed the fire. It was all I could do each weekend to cut enough wood to keep us warm until the next. I remember very early on standing back to admire my handiwork, a neatly stacked quarter cord of wood, and honestly thinking it would get us through the winter.

For Liz, the winter was just as tough as she struggled to keep house under what can generously be described as primitive conditions. During the week we worked side by side in a nursery pulling trees. Weekends she cooked, tried to keep the place warm, and helped me load and stack the firewood. We also spent a lot of time replacing boards where the cold air slipped through; clearing away tree roots when the ground wasn't frozen in preparation for a vegetable garden; and trying to isolate a persistent but evasive leak in the roof.

It was probably our roughest winter, and at times ten feet from the wood stove there was little difference between the inside and outside temperatures. Both the cooking oil and toothpaste froze solid, and the bucket of water we kept by the stove in case of fire was usually a block of ice by the time we awoke in the morning.

We made the smallest room in the back our "bedroom," covered the walls with $2 quilts from yard sales, and kept an electric heater going all night. Still, there were mornings when we awoke to find ceiling and walls covered with ice from the moisture which had condensed there overnight from our breathing.

We made it through that winter, eating as our budget would allow a lot of oatmeal and the occasional treat of cornbread supplied by our neighbors Doug and Terry.

Looking back, it was a hard winter, but we never really thought it was that bad at the time. I guess the combination of love and youth can really dull your senses.

With the Spring thaw we were ready. Apart from deluding ourselves into believing we would soon be self-sufficient with the food we imagined we'd raise in our rocky garden, and a couple of errant goats and chickens, our plans were entirely focused on starting work on the cabin.

In the fall, we had torn away an entire wing which had been tacked on to the house and unknowingly burnt more than a thousand feet of sound chestnut boards. The removal was necessary as the roof had caved in, but we didn't think ahead and realize we might use those boards, or that they might be of some value other than as embers.

Taken just after we started our restoration project, you can see the white-washed logs at left, while the right side of the one-room cabin is still hidden under clapboard siding.

Nevertheless, the house was down to four rooms, the living room/cabin, and the three rear "rooms" tacked in back of it. None of the logs were visible inside or out, and the place looked pretty rough in general, especially at the end where I had cut away the boards which held the half of the house we removed to the cabin room.

I don't consider myself a carpenter now, and to look at the work I did then working with what had become a very dull hand saw, well anyone who knew anything about building could have only shook their head in disgust. I don't blame them.

Uncovering a cabin can be very easy, so easy it lulls you into thinking the entire restoration process can be a simple project. Uncovering the logs is a lot of fun, too, almost like unwrapping a present as you'll never know what you'll find. It's almost like opening a time capsule: people slipped all sorts of treasures, and junk, behind the walls or between the logs of their cabins. And, there's no way to anticipate what mice or rats might have used to build their nests in the nooks and crannies of cabins like ours, where the logs and been covered for decades.

We learned that the secret to uncovering the logs, and not miss anything in the process, is to move slowly and cautiously. And until you can get a look at the entire log walls, you can't judge the condition of your cabin. Most log cabins survive without damage once covered up, but there's always that chance a weak log, or weakened wall because of unseen structural damage, could collapse when uncovered.

We also learned to proceed slowly so we could learn something about the history of the place. As you uncover logs, you're peeling back pages of history. Inside, layers of wall paper tell you something about the prosperity of the homeowners and the eras in which the house was redecorated. Outside, every modification which has been made to the cabin also tells you about previous owners, and their changing needs.

The types of nails used, even wooden pegs on a particularly old cabin, tell you something about the challenges each generation faced. There's also history in the boards used to cover or build onto the cabin, the way they are cut, the type of saw used, as well as the type of wood.

We wanted to learn all we could about our cabin, and were slow and deliberate in uncovering it. And we didn't throw away anything , even the varmint nests, until we had a chance to sift through each. Liz found an 1847 half dime which had been strung with a ribbon and once worn as a necklace in one of these nests.

All of that said, we could barely contain ourselves as we started tearing away the boards covering up our cabin. The only tools required were a hammer; crow or pry bar; nail puller and assorted screwdrivers;

even the type of tire iron that used to come with cars and has one end like a giant screwdriver.

Liz and I began working side by side near the old rock chimney, prying loose the clapboard from the top, one piece at a time. After a while, we were sliding the crowbar under three or four pieces and shaking the boards loose. In our eagerness to see what we were uncovering, we started working different sections at the same time. Liz continued working the side by the chimney, where she later found that half dime, while I worked the area around our front door.

We quickly discovered that in covering our cabin up, as with most cabins, some previous owner had tacked 1x4s, and the occasional 2x4, to the logs so they could hide the logs beneath the clapboard and end up with an even wall. Sometimes, they would hack away enough of the logs to make a straight wall; fortunately they didn't do that to ours, at least not on the outside.

Over the next week, as we worked the log cabin slowly emerged from within our clapboard house. As soon as we could see more a sizable section of logs it was apparent they had been painted white, repeatedly. This was "whitewash" we decided, and we could scrape thick flakes off with our fingernails. Needless to say, we were both a little disappointed at the thought of having a white log cabin, or having to strip all the logs. Later, we learned these repeated coats of whitewash were probably the best thing that could have been done for our cabin as they helped preserve the logs.

As soon as we started uncovering the logs, we began finding all these "relics" between the clapboard and log walls. Down through the years a succession of varmints built their nests there, and it seemed past owners may have placed some things there deliberately.(We didn't discover some of this stuff until later, when we took the time to go through boxes we filled as we cleaned away debris from between and around the logs).

It was quite a assortment of odds and ends: the nub of a home made pencil, bottle caps, countless buttons, spools of thread, a ball of twine, that half dime, the tag off a World War I gas mask. We also found some larger items including a mule bit, a fire heated piece of iron for putting pleats in skirts, a tobacco tin, an old axe head, and an ancient bottle of castor oil, some still pooled in the bottom.

And some of the boards we removed told a story of their own. We found a few pieces which had been once been part of signs. Another board was part of a packing crate shipped to a local store which we later learned operated in the 1800s. One board had the penciled receipt for some work done scrawled across it in an old fashioned script.

We were looking for some hint as to when the cabin had been covered up, and originally built. The previous owners had owned the house from the late 1920s until their heirs sold the place to John in 1977. By the time we were working on the cabin we had met Toy, the oldest son in that family, and his wife Morene, both well into their 70s. He told us they had never seen the logs, though knew of the one room cabin.

The only clue we found was a chad of newspaper stuck to one of the boards to which the clapboard had been nailed. Only the year was visible, but it clearly said 1911, so we knew the cabin had been covered then, at earliest.

Working together, it only took Liz and I about a week to expose the three exterior walls of the cabin (the back wall was still enclosed within the house, under layers of wall paper and drywall). Once the removal process was finished, we stepped back to admire our work before closely inspecting the entire cabin.

We had uncovered a 16x18 room on three sides. From the front it looked like any old cabin, with hand hewn logs joined neatly at the corners by tight dovetail notches. The entire cabin wore remnants of several coats of white wash. In many places this had flaked off to leave only a dull coat, while on other places it clung to most of the logs in thick slab which could be easily chipped away.

There was dust everywhere from the original mud chinking which filled the spaces between the logs. In some places it had crumbled away but most of it was still intact, some still covered with same whitewash which coated the logs.

"I guess we'll have to paint it all white," I said to Liz, as we stood wondering what to do next. She went and filled a bucket with some water, got a scrub brush and started working on a small area of one of the logs. To our surprise the whitewash just melted away, exposing the natural brown of the log beneath. As soon as the water touched the chinking though, it turned it to mud which dribbled down the face of the logs. We agreed that before we could clean the logs we would have to clean out the chinking.

Our cabin as we found it, hidden beneath the weatherboard, as John debates chigger prevention with Liz and Manny, our friend who helped with early demolition

Using screw drivers, paint scrapers, and the occasional knock of the hammer it only took a couple of hours to remove the chinking. But what a dusty mess! Here and there you could make out the finger marks and hand prints of whoever it was had daubed the mud in place.

In some of the tightest places, it was just that, mud, but it others they had filled the spaces with scraps of boards and rocks, whatever was handy, and packed the mud in around it.

We also found that some of the strongest chinking was actually a mixture of dirt with straw, embers and ashes from the fireplace, and even some animal hair. All of this gave the chinking a consistency almost like concrete, and it was in these sections we needed the hammer to remove it. I did leave one section of the chinking intact, at one corner of the uppermost logs, for posterity. (You can still see it in the upper corner of our family room today, if you know where to look).

Once the chinking was removed, Liz and I set out to give the logs a good scrubbing. A hose would have made this part of the project much easier. By then, we had progressed from hauling water from the spring in the woods to drawing it from the old well in the yard. Our well bucket only held two gallons at a time, so the slowest part of the project was drawing the water to fill our five gallon buckets. The combination of the whitewash and the mud dust quickly rendered each bucket of water useless for any more cleaning. So Liz did a lot of scrubbing while I spent the time cranking the well handle to keep the buckets filled.

As we worked on this part we quickly learned a few things that helped us work out a system for cleaning the logs. We had started working different sections at a time, one working low, and the other high. Soon, though, we saw our efforts ruined as the water from the upper logs trickled down the walls and messed up the freshly cleaned logs below.

From then on we systematically started cleaning the uppermost log, and worked our way completely across it, before working our way down to the next log.

And we ended up working from three buckets at a time: one for the scrubbing done with the brushes; the second for wiping away the dirt raised by the brush, and the third for a rinsing down the log with another rag in step three. Working this way, side by side, and in a couple of afternoons we had a clean cabin, ready for chinking.

Only as we scrubbed and cleaned the logs, did we discover how beneficial the repeated coats of whitewash had been for our cabin. If you've ever seen an old log cabin that's been exposed to the sun and elements, its usually a dull gray. Our logs, having been protected by years of whitewashing, still held their original color beneath the whitewash.

More than two weeks had elapsed since we first started working on the cabin, and our work had attracted a daily caravan of the curious who drove by to see how we were progressing. When Toy came by, he was able to point out the difference between the chestnut and poplar logs which had been used to build the cabin. I learned how to recognize each by the color and grain of the wood.

And it was then, while looking at the logs much as they must have appeared soon after they were put up, that I first appreciated the humble craft and determination which makes an authentic log cabin. The logs were of all different widths, in some places almost sitting on each other, in others separated by gaps as much as six inches wide.

All of them bore a series of hatch marks across their face, which I would later learn through experience, revealed where the man who built this cabin had hewn the squares faces of the logs from rounded trees. I spent a lot of time admiring the rustic simplicity of our pioneer cabin. From that day forward building one was something I wanted to experience myself.

But...back to our cabin. Once the logs were cleaned we checked each for signs of decay which might have escaped us. Using a screw driver we probed the wood for soft spots or rot. Overall, the logs were in great shape, although the chestnut had aged much better than the poplar. Some of the poplar was soft at its edges, and here and there we found spots of dry rot which we later chiseled out and filled in with wood filler.

We were extremely fortunate, for in the entire cabin we found only one small log which had to be replaced: a two foot piece which set between the front door and window. It was just a matter of knocking it loose, cutting the edges of the nails, then slipping a new section in its place.

For its replacement, I cut a piece from what I now recognized as a chestnut log which had been used as a floor joist in the section of the house we had torn down.

I think it was Doug who suggested we treat the logs with linseed oil to bring out the grain. We took his suggestion and improvised, combining it and Thompson's Water Seal, to bring out the grain and offer some protection. We painted this on every log, all around.

The real estate agent who had sold John the property, also advised that we paint the house inside and out with chlordane to prevent any chance of insect infestation. I fault my own ignorance at the time for following his advice, and later regretted even the thought of doing this. Fortunately I only applied this to the outside of the logs. Had we painted the inside, I don't think we would have remained in the cabin once we learned about the dangers of this poison.

After the logs were treated and dried, we set to work on the chinking. For this we used ready mix Quickcrete mortar, which has held up surprisingly well. Most of it is still in place nearly 25 years later.

As we started to chink, we quickly learned why the old timers would fill the gaps between the logs with whatever was at hand. It takes a tremendous amount of "mud" to fill those spaces between logs, and without something else to hold it, it can just slough off the log.

Observers recommended hammering nails in place, stretching chicken wire, insulation, screen, and we tried them all. But ultimately our experience turned us back to what the pioneers used to hold their chinking: scraps of boards and rocks, tapped into place for a tight fit. It's an inexpensive and practical solution, although a time consuming job.

Chinking takes a lot of time; in fact, it's the part of working on cabins I enjoy the least. You can tell by looking at my work. Liz is much more methodical about it, and she works the mud to a nice even finish.

Mine looks like what it is—thrown into place. She started using a trowel and scraper to smooth the surface, but later just worked it in with her hands for a more authentic look.

After only one day of working with the mortar in our bare hands we realized it can take a toll on your skin. Our hands were rough, red and chafed. From then on we always wore plastic gloves when handling our "mud."

Within a few weeks, all the logs were chinked, the first phase of our restoration complete. It would be 8 years, and four children later—the oldest two born at home in our log cabin—before we would resume the restoration of the one room cabin.

We lived in the cabin after uncovering and restoring the outside for another three years. Our first two sons, Marcus in 1979 and Luke in 1981, were born at home, delivered by a midwife. Because we had toddlers under foot and were using the cabin as our living room we did not want to deal with all the filth or chaos that an inside restoration would create. So the interior walls remained covered under the layers of drywall and wallpaper as we had first found them as we started on our family.

By 1981 I was working as a reporter for the local paper, making barely above minimum wage. With our family now grown to include two sons, Liz and I reluctantly decided to move back to NY where we could make more money. By then we had purchased the entire farm—22 acres and the old cabin—from her father, who had given up his plans of retiring to the country.

So in November 1981 we reluctantly left vowing to return. For the next three years we spent every summer's vacation in Tennessee. All the while the rest of the year, we talked of eventually moving back to Tennessee and finishing the work on our cabin. We had decided we wanted to raise our children in the country.

During those three years we added two more to our brood, Sarah and Sean. When at last we finally did pull up stakes, deciding if we didn't make the chance to live our dreams we never would, we arrived back home in July 1985 with a dire need for more space.

We managed to live in the cramped quarters of the little cabin and three rooms through the winter of 1985, planning a major renovation for the spring. Our intent was to do as much of the work ourselves as possible, as our financial situation dictated we cut every possible corner. We planned to tear away everything but the original cabin and rebuild above and around that.

That winter I literally carved the space for our the extra rooms, using a pick axe and shovel to cut away the bluff surrounding the cabin. Working nights and weekends I made a space for our expanded house one wheelbarrow at a time.

When we started work that spring it took more than a month of working together, evenings and weekends to tear down the old house and prepare the building site. The fact that we had four little eager helpers only slowed the work at times but made it a real family project.

For a few weeks after preparing the site but before beginning work on the addition, the cabin stood as it had originally been built, a simple one room house at the edge of the woods. It was a humbling experience, one which made us appreciate the sacrifice and determination of the pioneers.

For the months Liz and I were doing this work, with four kids always and everywhere under toe, we had a real taste of the pioneer experience: all six of us lived within the confines of that one room cabin. We did have an electric stove and sink in the front yard where Liz cooked and cleaned for this period, but the 16x18 cabin was our house. Each corner of that room cabin served a different purpose, much as it must have been for the pioneers.

We had our dining table and food cabinet in one corner; a bed for the all four kids in the other, a sleeper couch for ourselves against the opposite wall, and on the opposite corner a TV, couch and radio. Our cat even managed to sneak in and find a place to have a litter of kittens under the bed.

Thankfully we had wisely waited until spring to begin the work. Once the kids were asleep, Liz and I spent most of the evenings in the yard, discussing and refining our plans, until we were ready for rest.

Since we planned to add a story above the cabin, and rebuild around it, we had removed the drywall and wallpaper which had been used to cover the exterior wall of the cabin long ago, when the small rooms were added. This back wall was in as good shape as the others except for the exposed surface of some logs. Whenever those rooms had been added, someone had roughly hacked at them to nail 2x4s in place to make a straight wall to hold the drywall.

Other than that, though, this back wall and its logs were sound. And now that the exterior of the cabin was completely exposed, as it had been built, we could see where the roof rafters had originally set. The third log from the top had a series of six crescents cut along its upper surface, evenly spaced, between it and the next log.

I later learned in a museum, that as our cabin was originally built this would have been the top log. Those semi-circles were where the log rafters of the original roof rested, supporting the slab or shake roof. The two logs above it had been added when they needed more room added a loft. By then they had access to mill-cut rafters and rough mill-cut chestnut boards used as the floor of the loft.

Looking back, with what we know now it should have been easily apparent to us that the loft was not part of the original cabin. Whereas the rest of the logs were hewn flat on two sides, facing in and out, the upper logs had been hewn square. What's more these logs were joined at the corners using a flat saddle notch, held in place with locust pegs, whereas all the other logs featured dovetail notches.

The uppermost of these squared logs was about two inches wider on the interior wall than all logs below it. Small poplar trees which serve as the pole rafters for the roof were nailed to it. On these were nailed, rough cut lathing of various widths, topped with a shake roof. The present tin roof had been constructed sometime in the 1920s, but we did find a couple of the original oak shakes where had been left among the lathe.

Since we knew next to nothing about carpentry we hired a friend Jim and his helper Reece to frame the rooms we wanted to add around and above the original one room cabin. Liz and I did as much of the work as we could to save money, often working late into the night after the carpenters left. Liz also studied up and supervised me on running the electrical wiring for our new house.

We learned a great deal that spring and summer which has served us well in all subsequent building projects.

After two more months, we were all in new rooms, all four kids sharing one room with double bunk beds, and Liz and I moving into the loft room above the original cabin. We had a kitchen, and a middle room which functioned as a den. We still had our shower and outhouse out back, but we felt like royalty with all the clean space we added.

And this gave us the space to resume the restoration of the cabin interior without disrupting our lives. That project presented more filth than anything else. Beneath the drywall I think we counted 11 different layers of wallpaper, and at least one layer of newspapers pasted to the logs from the mid 1800s. There was some chinking still intact, but most of it had been knocked loose years earlier when we chinked the outside of the cabin.

Once we cleaned up all the dust and crumbled mud, it was just a matter of scrubbing down the logs, treating them with linseed oil and applying the chinking. We were expecting to find a treasure trove, but only found more odds and ends between the wallpaper and logs. This time the haul included more buttons, a couple of mummified mice, a ball of string, a homemade pipe with a tiny twig bowl, a twig toy, and a piece of an old sign listing the prices of vegetables.

We also discovered some very large cut nails tacked into the logs which had apparently been used to hang pots or what have you around the fireplace. We left those intact as a reminder of how the room had once been used.

Between the cleaning and chinking it probably took another month to finish up the work. As a memento I had each of us press our thumbprints into the last bit of mud by the doorway. Our four oldest kids, grown now, can one day marvel when they compare how small their thumb prints were.

Because we had built a staircase to get upstairs to our room, I had actually boxed in one corner of the original cabin, which remains covered for now. I'm sure we'll get around to fully completing the restoration one day. At least that's the plan.

If you were to walk up to the front of our cabin today, under the porch awning, it looks much like it must have when first built in the 1800s . The porch runs the front length of it, and around the side to the limestone chimney. Step inside the front door, you step into a room with the original mantle and low ceiling. The crooked floor, irregular walls and uneven gaps between the logs all contribute for the personality and feel of an authentic cabin.

You can catch glimpses of the cut stones which were used to build the fireplace behind the mantle. The original back doorway is still directly opposite the front door, and there's another doorway, added later on the right, where a window once looked out on the dirt road, the wilderness and valley beyond.

Step through either doorway and you're outside the cabin: it is the centerpiece of the log home we've created around it, retaining all its rustic charm, always inviting an escape into simplicity from the confines and confusion of a modern house filled with six children.

Our hearts are in the cabin, and the valley that first inspired kindred spirits to make their home here in the hills, on this spot along this road.

For the next two years we remained thrilled with our new space. But as the kids grew in size and stature we recognized we hadn't built enough onto our home, and we started thinking of ways we could expand.

It would give me the opportunity to experience first hand the challenges the pioneers faced, and the satisfaction of transforming trees standing in the woods into a rustic home, working with many of the same hand tools.

2

Methods for Uncovering and Restoring An Old Log Cabin

Restoring an old cabin may be the easiest way to get an authentic log cabin. And, there's certainly a lot of fun to finding and uncovering that diamond in the rough and transforming it into the jewel you will proudly call your own.

As in the rest of the book, this second part of the chapter describes the methods we used, as well as insights drawn from our experiences.

Here, the focus is on finding, uncovering and restoring a vintage log cabin.

FINDING YOUR CABIN

If you want to find an authentic log cabin you have to know where to look. Your best bet: head for the countryside. Depending on where you live and how long ago it was first settled, you may not need to look too hard to find a cabin just waiting for someone to come along and treat it to some much needed TLC. They are out there, but their ranks are thinning.

Where we live in Tennessee an afternoon's ride along the backroads leads past some old cabin still standing, and others sliding toward oblivion. There's fewer left each year, but they're still here and there.

More likely, though, you'll have better luck searching inside other structures to find that authentic cabin. The log cabin was often the first dwelling people put up when they settled an area. It offered a relatively quick and extremely affordable shelter using the materials at hand. And the pioneer couple, or a man working alone, could have a crude cabin ready to move in within days or weeks, depending on how much help was at hand.

Some of these earliest cabins were little more than enclosed lean-tos, built of round trees or saplings. Many of the earliest had dirt floors which made them extremely prone to both insect infestation and decay, so few survived. Later arrivals, or those who worked in the warmer weather and had some time for their project, had more options in the way they built their cabins. They may have been able to hew or square the logs, build a floor, even of rough planks or half logs, and take the time to construct the cabin so it set up off the ground.

Most surviving early cabins were built this way: raised off the ground, supported at the corners on stone pilings or piers, with half hewn logs for floor joists. If there was a saw mill nearby they might have milled lumber for the floor, otherwise they would hew or cut rough slabs from logs for use as flooring.

The typical one room cabin was usually under 20x20 feet, with a door, a window or two to let in some light, and a fireplace at one end. Some of the earliest fireplaces were built of mud and sticks, later replaced with cut stone.

Though many people started in simple homes like this, they often associated the cabin with hard living and poverty. As the need for space grew, and the family became more established, they might start modifying that simple cabin by adding a loft or tacking rooms on the back. Eventually, some subsequent generation would completely cover the original cabin under a layer of clapboard or weather board outside, and inside behind a straight wall of planks to which they could affix wallpaper. Some initially covered interior logs under a layer of newspaper, and later paste wallpaper directly to it.

Methods for Uncovering and Restoring An Old Log Cabin

To look at one of these older homes, and even walk through it, you might not guess there was a log cabin or log home hidden within the walls. But there's a couple of places you can look for clues.

Start by asking around if anyone knows the history of the home. Usually if there's a log cabin there, it's something which has been passed down generation to generation, and people who grew up in the area will be aware of it.

Don't take anyone's word alone though.

Here's some signs which will tell you there is, indeed, a cabin hidden within:

The logs used to build cabins were generally at least six inches thick, compared to walls of about 4 inches in the modern home. Look along the window sills and door jambs; the log wall is usually thicker than other types of construction.

Some "Survivors": Four examples of original log cabins once standing alone or inside other homes awaiting restoration

The shape of the room, as well as the height of the ceiling can also be good indicators. Log cabins tend to be square to slightly rectangular. If a loft was added to the original cabin, the ceiling is usually lower than the eight foot you expect in modern homes. If there's an attic above the room you suspect may hide a cabin, be sure and check there as you may be able to see sections or ends of the logs there.

Next look under the room. There, you may glimpse logs which have been flattened on one side to serve as floor joists. Look to the exterior wall where these joists rest on the sill and you may also be able to see the bottom of the lowest "course" or level of logs there.

Usually a one room cabin has a fireplace or chimney situated in one wall. The room with the fireplace is the most likely candidate in any older home. Examine walls and corners for anywhere you may catch a glimpse inside the walls. As in our house, you may get your first look at the logs where the wall meets the chimney.

You've got to get a good look at a section of wall before you can tell if there really is a cabin there. For this you're going to have to pry some boards or paneling loose. You can learn the most from corners, where the two walls intersect. Look for a spot which is out of the way, under the staircase if the room has one. Outside, look first for weather board or siding already loose.

When you can peer behind those boards, look for at least two logs, and the space between them which may or may not be filled with chinking. Once you see that it's a good sign you've got a cabin. Still you won't know it's condition, and whether or not it can be salvaged, until you start your restoration by uncovering the cabin

Tools

Unless you discover your cabin requires extensive restoration or replacement of logs, you won't need any specialized tools for this work. Over time, though you will use a surprising variety tools for all that's involved in the restoration project.

You'll find information on the complete range of tools used in all aspects of a log cabin projects in the Tools appendix at the back of the book. (*You may want to review that entire section as you explore each of the "methods" sections for the various types of log cabin projects.*)

UNCOVERING THE CABIN

Before you begin uncovering a cabin you need to determine how the work may affect the physical integrity of the house and any surrounding and supporting structures.

In particular, you want to determine if any boards or posts you plan to remove will weaken the support of the walls, ceiling or roof which have been added to the original structure. You may need to brace these before embarking on your work.

New posts may need to be added further out from the walls covering the cabin, or you may need to brace a wall so it will continue to stand when it's no longer attached to the cabin. Be thorough and overcautious. Before the work proceeds you want to take steps to assure you can uncover the cabin walls without weakening the surrounding structure.

As you inspect the walls covering the cabin, look for any signs of a persistent leak or water damage as these could be problem spots which warrant closer inspection, and support, before you can safely proceed with the work.

When uncovering a cabin the debris quickly piles up. You will need an area away from the cabin to store what you remove so you can work safely. Some boards are bound to break and be rendered useless as you remove them; others you'll want to set aside for future use. Plan ahead on where you stack these boards. If you are uncovering a cabin inside a home, decide on the path you'll use to carry off the debris, or through which window you'll toss the wood to the outside.

As you start, inspect the boards, paneling or drywall which covers the cabin for vertical and horizontal rows of nails or screws. At these points the boards are attached to studs or cabin itself. As you begin removing the boards uncovering the cabin you want to work the crow bar as close to these nail lines as possible. Work them carefully along these nails and you'll end up with many reusable boards.

Ideally, there's enough gap between the boards to slip the tip of whatever tool you're using to remove the boards, and press against adjacent boards for leverage. If you find you have to slip your tool behind the wall and press against the cabin, slip a scrap of board at the contact point between the bar and logs otherwise you may leave unwanted dents in the wood.

It's safest to start uncovering the top logs and work your way down. If the cabin has been covered with lapped weather board or clapboard outside, this is the only way you can save the boards. Loosen the boards at the corners first. Before you can do that, you may also need to remove a vertical board which covers the edges of the boards where the corners meet.

At any rate, loosen the boards at one corner, then work your bar toward the middle, and on to the opposite edge. Be careful not to let the board fall on you as you work; if you have a helper, get them to hold the board, and slide it out of your way once it's loosened. Working in this way, concentrate on one wall, and one section of it, at a time.

You can also remove the boards by pulling each nail as you go, but this can slow the process down. It takes time to position your tool properly to pull a nail and then work it loose. Working with a crowbar, you may even be able to remove several boards of lapped siding at a time, and pull the nails holding them together later.

If your work uncovers any sign of weakness in the wall, such as loose or leaning logs, stop until you can make sure these logs are safely secured. Occasionally some logs will rot or were already showing signs of decay when covered. If you aren't cautious, there's a real danger these could collapse when uncovered.

Usually these problems are most likely to be found on the sides subject to prevailing winds and rain, or beneath any persistent leaks in the roof. The wall surrounding the chimney may also be prone to damage, as rain often runs down the chimney rock and onto to the logs

Remember as you uncover the logs to have a box or bucket nearby to collect anything you find, including rodent nests (if you want to go through that filth later). Be alert for any scraps of paper, or boards which may tell you something about the history of the cabin, or when it was covered up.

In particular pay attention to where you encounter any cut or square nails, even wooden pegs, as these can help you date the structure.

As you work to remove the boards around windows and doors take extra precautions. Often the original windows were replaced or new windows added as the cabin was expanded or covered. Sometimes the new windows were built onto the frames which were part of the original building.

After you've uncovered the entire cabin you should sweep the work area clean for the next part of the project: cleaning the logs in preparation for applying some preservative or finish, and chinking. Take the time to look over what you are sweeping away or throwing out, as you may have missed something of interest as you uncovered the logs.

CLEANING THE WALLS

Before you can begin cleaning the walls you need to remove the chinking from between the logs. This can be an easy but messy job. You may be able to lift some of the chinking out. Otherwise you may need a screwdriver or hammer to knock the chinking loose. Pay attention to what the chinking is made of, and what it includes as this may hint at some of its history. Also, be alert to what else may tumble out as you remove the chinking , especially from spaces near corners, windows and doors.

The front of the cabin as we found it. The logs all bore a thick coat of white-wash. In the lower photo the boards to which the clapboard siding was nailed are still in place.

Logs which have been covered for decades need a good scrubbing before you can apply any protection or preservative. Once you determine the logs are sound you can use a hose or pressure washer for the preliminary cleaning. If neither is available, good old fashioned elbow grease works fine, it just takes a little longer.

If the cabin is inside another building you should take precautions to ensure the cleaning won't soil or damage the rest of the building, or furnishings. Seal the inner walls with plastic so the water won't find its way inside the house and cause damage there.

Whether or not you can use a hose, the logs will probably need a good scrubbing with a stiff bristle brush. Use a standard bristle brush and a nylon brush with handle, dipped in water.

Unless you regularly mop up or wipe away the dirt you're raising with the brush you'll be left with logs which are still dirty. As I mentioned earlier, we kept two additional buckets of water on hand, each loaded with rags soaking in the water. The first we used to wipe off the dirt raised with the scrub brush, the second to completely wipe each log clean. Regularly changing the water in all the buckets is key to getting the logs clean.

We didn't use any type of soap and found the water alone did the job. You can add soap to your water if you wish. If you do, use some mild wood cleaner like Murphy's Oil Soap.

If your earlier inspection detected any areas of decaying logs or rot, you want to be careful not to soak these areas when cleaning your logs as this could hasten the progress of the decay.

In the unlikely event your logs have been painted over in decades past you may need to resort to a more radical cleaning method to restore their original color. (Since I never had this problem I don't want to offer any advice on options I haven't had occasion to use. I can say I would hesitate to employ any type of solvent or paint stripper but only as a last resort.)

Sanding the logs by hand seems prohibitive as an alternative as the unevenness of the logs, and the presence of cut marks from the ax or adze will make it virtually impossible to hit all areas of the log which need to be sanded.

I've also heard you can use a sand blaster to work off the paint, but I would be concerned about damaging the logs if I didn't know how to use one. An alternative would be to use something less abrasive than sand for this process. I've heard of people using both ground corn cobs and black walnut hulls as a substitute. If you are considering this, I'd recommend you have a look at some project where this method has been used to strip wood before deciding if it's right for your cabin.

Whatever method you use to clean your logs, make sure they are thoroughly dry and free of dust before you apply any treatment.

REPAIRING/REPLACING LOGS

Your inspection may reveal section of logs, or entire logs which must be replaced before work can continue. If you find areas of dry rot or decay, this should be removed, as surrounding areas of wood could become infected.

Probe the affected area with a screwdriver to determine the extent of the damage. Then, use a hammer and chisel to remove all the damaged wood, leaving a border of sound wood.

Once the damaged wood is removed, I suggest you use a wood hardener before you fill in the affected area with wood filler or wood putty.

I've also used a MinWax product, "High performance Wood Hardener" as a treatment for decayed or dry rotted areas of logs which I could not remove. This hardener penetrates the softwood, and hardens within a few hours to restore strength to the wood and make it impermeable to further damage from water. It's expensive, but well worth the cost of preserving the log.

Once it's dry, you can fill the surrounding area with wood filler. For this we've used the powder you mix yourself. It's relatively inexpensive, long lasting and holds up well. Once the filler dries it will take an artist's eye to match the color to the surrounding log.

When the damage is more extensive than can be solved with wood hardener and wood filler, you may have no choice but removing and replacing a section of the log. It's not that complicated: simply cut the old log from the wall, and slip a new piece into place. If you must replace a section of one or more logs, weigh the merits of replacing the infected area with a small window. This may mean you'll have to remove sections of adjacent logs, but cabins can always use more windows and more light.

Of course, you'll want to find a piece of log which closely matches what you need to replace. If possible, salvage a usable piece from materials removed as you uncovered the cabin. If you can't, you can hew out a piece from the appropriate type of timber(see Chapter 2, Part II).

Or, you can go to a local saw mill, and have them square off a piece of log of suitable type. When possible, use the same type of wood as the log you're replacing. If you do use a saw cut piece of log, you can rough it up with a hand axe later to give it the hand-hewn appearance if it's that type of cabin.

If you cut a fresh log for the replacement piece, remember that over time green wood will shrink in width but not length. If working with a "green" fresh cut log which hasn't cured, use a piece that's slightly wider than the log you're replacing so it will dry to about the same size. When it shrinks, the chinking will cover the difference.

Before you can replace the section, you must entirely remove the decayed portion of the old log. First, support the log containing this section, as well as those immediately above and below. Build a frame large enough to isolate the section of log you are replacing. Nail this frame to the logs before you proceed. Be sure to remove a section that contains the entire decayed area and wood immediately surrounding it.

When inserting a new section into an existing log, you can use a simple lap notch at both ends to hold the piece you're adding in place. Again, before you cut into the log, be sure to support both ends of the log you're repairing.

Once you've removed the section of log, cut and remove a rectangle at least four inches long and three inches high from both ends of the log you're repairing, and on the opposite ends of the piece you'll be inserting. Slip the piece into place, and "toenail" the logs together at an angle to hold the piece in place.

Replacing an entire log can be handled the same way, as long as you can cut the log so the notched end remains intact. However, if the log and notch must be removed the process becomes very complicated and dangerous. Before you can remove the log you must brace and support all logs comprising the upper section of the cabin. Unless you dismantle the upper portion of the cabin, you will need to raise and support the upper logs before you can slip a new, notched log in place. Investigate all other options before embarking on such a drastic measure.

TREATING THE LOGS

If you plan to treat the logs with any type of preservative or stain, do it before you begin work on chinking. If you wait until after the chinking is in place, you'll cover the "mud" as well as the logs, and that could leave ugly stains on the chinking.

The pioneers didn't apply preservatives to their cabins because they weren't available to them. Many log cabins have not survived because of this, falling prey to insect infestation, dry rot or water damage down through the years. So whether it's a cabin built today or 150 years ago, it's a good idea to apply some sort of treatment to the logs.

In our case we combined linseed oil, with Thompson's water seal. (I'm not endorsing Thompson's. It's what I found available and worked fine. I'd use it again and again.)

As far as spraying the logs with an insecticide, I am against it unless absolutely necessary, especially on interior walls. I believe anything which kills one life form cannot be entirely harmless to others.

That said, however, there are instances where you will have no choice but use an insecticide on the logs, or surrender them to termites. If you have to treat them for insect infestation, I strongly urge that you research your options, and the associated dangers of the chemicals, before applying any log treatment.

If you do decide to use a chemical protectorant or insecticide, stay out of the cabin for as long as you can after the treatment is applied. I'm talking days or weeks, not hours. I know of one person whose lungs were scarred by returning to a cabin without allowing sufficient time for noxious fumes to dissipate.

Also, before using any sort of preservative, stain or oil for your logs make sure you try it on a sample board or section of log first. You may be surprised at how some of these cure in their final form. One hardware dealer sold me a copper-based preservative for my cabin with the assurance that it would dry clear. I didn't want it once I read the warnings about toxicity on the label, but by then I had already painted a sample board which dried to a permanent pale green.

There are many choices in waterproofing and wood preservatives which will produce adequate results.

Here's a quick run down of what we've used and would recommend:

Boiled linseed oil: Darkens logs but brings out the grain nicely. Apply when warm and it will go much further. Has a distinctive smell which lingers for a long time, which you may not like. Also, it leaves a slightly oily finish to the logs which can become a magnet to dust.

WaterSeal/waterproof. : Easily applied, these wood protectors , once dry, make the logs water repellent. Dries quickly and serves for years. Good for use as an external treatment.

Polyurethane finish: Available in flat, satin and gloss finishes, in a variety of wood colors. Easy to apply, slower drying than water seal, but provides a protective coat while helping to bring the grain out of the wood. I used a satin finish for interior logs on one old cabin and was quite pleased with the results. It took a couple of days before they were dry, and several more before that odor had faded enough for us to remain in the room comfortably for any length of time

To expose the interior walls of our cabin we had to remove drywall then layers and layers of wallpaper.

Preparing To Chink

Once your log walls are cleaned, treated and repaired, it's time for chinking. This is the most frustrating part of working with logs as the work progresses so slowly, especially when doing it alone. Accept any offer of help, in fact, you may want to host a "chinking party" to get this part of the work done as quickly as possible. Many hands make for short labor.

Expect to be surprised at the amount of mud required to fill the spaces between logs, so do all you can to reduce your need, and strengthen the walls at the same time. Rocks or pieces of boards slipped between the logs will lighten the load on the wall and cut down on the amount of chinking required. They also give the mud something to bind to once you start filling the cracks.

Collect rocks of different sizes as well as scraps of boards, and make sure none are wider than the width of your logs. Before you mix your mud, push or hammer these rocks or boards into place, filling as much of the gap as possible between the logs.

As for the mud, we started using Quickcrete mortar mix, and it took many 80 pound bags to complete the chinking of the outside of our cabin. For subsequent projects, I asked the owner of the local cement factory what he'd recommend for chinking. He suggested a mix 1 part mortar mix, 1 part Portland cement and 5 parts sand. That's a rough approximation; you'll have to adjust the amount of sand and water used to get the desired consistency of the mud.

When it's right it should have the consistency of mashed potatoes. The most common mistake made when learning to mix mud is to make it too dry, with either too much sand or not enough water. Dry chinking is prone to crumbling. You'll know when the mud is too wet because you won't be able to get it to stick to anything. It may take a couple of batches before you get a "feel" for mixing the right mud.

When we first started working on this cabin, I mixed the Quickcrete by the bag in a wheelbarrow, following the directions for adding water on the bag. Working the water into the mix back and forth with a hoe proved a very tiring process, especially when you consider the amount of time spent to mix the mud with the relatively short time required to apply it. I later tried to get more from my effort by mixing the two or three bags at a time in a kiddie pool. I got more done at a time, but this proved just as tiring.

Finally, I purchased an electric cement mixer at a home repair center. It seemed at an extravagance at the time, something like $200. I can't begin to estimate the number of back breaking hours it has saved me on my various projects.

If you anticipate using a cement mixer in the future I recommend you buy one, if not; rent one for the days you plan to do your chinking, and make sure you have some extra help available. That way you can apply the chinking as fast as you can mix it.

Applying The Chinking

Chinking a cabin can be a very wasteful process. No matter how careful you are, some of the mud will end up on the floor. We learned this early on. Now before we start working on any wall, we spread a layer of plastic or cardboard on the ground right up against the bottom log. Then we'd frequently gather up the mud collected there to use as chinking.

Since the logs in an authentic cabin come in varied sizes, with varied gaps between them, you'll probably use a combination of trowels of different sizes as well as your hands to force the chinking between the logs.

Start by putting as much mud as you can comfortably handle on your mudboard. For this, use a scrap of 1/2-inch plywood you can comfortably hold. I keep my trowels in the mud, and rubber gloves nearby in case I need to work the mud between the logs by hand.

Start with the top logs and work across, from one of the end of the log to the other. Place the flat edge of your mud board about an inch below the bottom of the log you want to chink. Slide the mud from the board into the gap separating the logs.

If you're working with a trowel, you'll have to push a gob of mud there, and then use the trowel to spread it evenly. In tight places you'll find it works best to apply the mud by hand.

When you've done a good job filling in the spaces with rocks and scraps of wood the mud will settle into the spaces between them. Inevitably some of the mud will sag a bit away from the upper log. In that case return to that spot a little later, after the mud has dried a bit, and apply new mud on top of it.

If you plan to reuse the mud which has fallen onto your tarp or cardboard, don't let it sit too long or it will dry out and not be usable. When your mud dries out a bit, you can work a small amount of water into it to restore the desired consistency.

After the mud has dried to the touch, usually in a couple of hours, it's a good idea you brush the chinking with a strong hard bristled brush. Just brush back and forth along with enough pressure to rough up the surface of the chinking. I learned this trick from a brick mason. He explained that if you don't do this the mud can dry to a shiny smooth finish, something you may not want for your cabin

One more thing: when working with concrete always remember to thoroughly clean your tools with water, especially your mixing tray or cement mixer, as soon as you are through using them. Otherwise you'll have a hard time removing the dried concrete.

After a couple of hours, take a wet rag and go back over the logs to wipe away the white dust or spots left by the concrete, before it leaves a permanent mark.

MODERNIZING THE CABIN

Unless you're looking for that "primitive" experience, you'll probably want to add some "modern touches" to any log cabin…simple things we take for granted like electrical wiring or plumbing, finished walls and floors, energy efficient windows and skylights, insulation, etc.

These are briefly discussed in the appendix Adapting The New following the Tools section at the back of the book.

Again, you may want to read through this section before you finalize the plans for your cabin restoration or building project.

PART II

The Hand-hewn Log Cabin

1

A Hand-Hewn Addition, The Old Fashioned Way

Imagine building a log cabin yourself, working just as the pioneers did.

I thought it would be fun, but my fantasy never allowed for the amount of work involved.

From time to time, when I studied the original logs and the tightly done corner notches of our one room cabin, I wondered about the man who had built it, what it had been like, working with only a few hand tools at the forest's edge, to make a home for his family from the trees which grew there.

It was something I thought would be both fun and challenging, and our circumstance gave me the opportunity.

After Liz and I finished restoring our cabin, and adding three new rooms behind and above it in 1986, we were thrilled with the space. It seemed a mansion compared with the cramped quarters we started with. Even though our four children,—Marcus, Luke, Sarah and Sean, all under seven—shared a room and a pair of bunk beds, we thought ourselves rich in space…at least for a while.

But as the children grew they needed their own rooms. By early 1989, we were trying to figure out how to expand our home even further. We knew we wanted to build with logs somehow, so not to detract from the rustic simplicity of our original log cabin, which served as our living room.

First we explored buying a log kit, and building it onto our cabin. For us, with limited finances, the cost seemed prohibitive. And, there was another issue: in all the log kits we considered there was a "sameness" to the logs. The kits featured logs of uniform size and width, with the same gap between every log. They looked to us like striped houses.

We knew from our own cabin, and others we had explored in the surrounding area, that no two "authentic" log cabins are alike. The pioneers used whatever materials were at hand, and that meant trees with twists and turns, knots and varied widths. To truly match our cabin as best we could, and retain its personality, we needed the same variety.

But it would take more than a random selection of trees to create a match. Ours was a hand-hewn cabin, joined at the corners with dovetail notches.

Nothing less would do.

So, ready or not, here was my opportunity to experience building a hand-hewn log cabin, or actually 3/4s of one. Realistically, at the time I had no alternative as I figured building the walls of logs would mean substantial savings on the total cost of the project.

I'd wondered how one man, working alone in the woods, could do it, and what challenges he faced. Now, I would learn by doing, limiting myself as much as practical to the same tools available to the pioneers.

I made a point of telling anyone who cared to listen of my plans. Years earlier I learned that if you really want to be held accountable, let others in on what you intend to do. They will ask, and serve as a constant reminder, of how well you are progressing toward your goal.

Liz and I worked on a number of different floor plans, and finally decided a 26x26 foot addition to what we had already built would give us all the room we needed. The plans entailed a laundry room, bathroom, four small bedrooms, and a large family room. (In retrospect I wish we'd made the space larger; one thing I've rediscovered repeatedly is to build more space than you think you need and you'll find a way to fill it.)

At any rate, while Liz refined the floor plans, I focused on an education by the book on building a log cabin. I'd studied many log building since we first uncovered ours, and had actually watched a friend Jim square off or "hew" a log with an ax for a home he was building. My knowledge amounted to little more than that: what you can learn, watching from the shade, as someone else works.

I also read through a lot of "how-to" books, and found what I really needed in the few pages of a section on log cabins in the original *Foxfire* book. This gave me an idea of how it could be done. And I figured if some pioneer could head off in the wilderness with nothing but the idea and a few tools, I could do it too.

I knew from studying cabins and what I read that the first cabins in the part of Tennessee where we live were built from whatever logs were growing on or near the homesite. Soon, though, the pioneers developed a preference for a combination of poplar and chestnut logs. Both trees grew tall and straight, are relatively light and easily worked.

Chestnut logs and lumber were especially resistant to insect infestation and decay. The trees themselves proved susceptible to the chestnut blight, a fungal disease imported from Asia in the early 1900s, however. By mid century, it had all but erased the American chestnut from the forest.

The poplar, or the tulip tree as it's also called, is also light and known for growing tall and straight, making it ideal for log home construction. Poplar's major drawback though, is that it can quickly rot if left in contact with the ground, or exposed to water or moisture for prolonged periods. But unlike the chestnut, poplars, still grow in abundance in the steep forests which have reclaimed hill farms in our area. We had plenty of trees to use for our cabin addition.

It would take tools to transform those trees into logs for a home, and I needed to locate the traditional hand tools: broadaxe, broad hatchet, adze, assorted saws, chisels, and something called a "cant" hook, or timberjack, for moving logs. I also planned on using some tools which the pioneers may not have had, such as a level and chalk line, block and tackle, and a chain saw.

I did locate sources for modern reproductions of the traditional tools, but at costs well beyond my budget. I'd seen some of these tools in flea markets and antique stores down through the years, but now that I needed them I couldn't find them anywhere. I asked around and finally bought two adzes, three broadaxes, and a broad hatchet from Mike, a former hippie who had outgrown his interest in log cabins. He generously sold me the tools for only $10 each the night before Thanksgiving in 1988. Now that I had the tools, I could turn to the logs.

Our plans called for an addition that would measure 26 x26 feet on the inside. Since we were building on to our existing home, I only needed enough logs to build three walls nine feet high, counting the bottom or sill log. For the two long walls I estimated I would need logs 28 feet long. Because of the layout of the old house, and how we were adding onto it, logs for the third wall needed to be only 18 feet long. Figuring I wanted logs with an average face of 10 inches, with two inches of chinking in between, I allowed for one log for each foot of wall height. So I figured I need to cut 10 logs for each side, and then allowed for an extra four logs, just in case.

That seemed like a simple request until I started walking through the woods on an initial search for trees I might use. I found plenty of the right length, but rarely could I find trees that gave the straight run, or nearly straight run, I needed. There were a few, but most featured a twist, bend or limb which made it impractical to use them for the required length.

So Liz and I returned to our plans and tried to improvise. From what I had learned all four walls of a log cabin were usually built first, and then the doors and windows cut from the standing logs. We figured if we built the doors and windows into the walls as we raised them we wouldn't need near as many full length logs.

So we went to a salvage dealer and bought the windows and doors we would use. We bought a large picture window for the family room, a 3x5 window and smaller windows for each of the other rooms. We measured the outdoor dimensions of each, as well as the casement for a 36-inch front door, then stored the door and windows in our barn until we would be ready to use them.

Using those dimensions, we planned the placement of the windows and doors in each of the walls of the log addition. This gave us an idea of how many logs, and what lengths I would actually use in raising each wall. Based on these estimates, I needed only two 30 foot logs for the front wall, four for the side wall, and the five 18 foot logs for the back wall. All the rest could be cut from smaller sections of trees. I figured I could get by cutting 30 trees of varied lengths, each with at least 20 feet of usable logs.

Now that I knew almost exactly what I would need, I selectively decided where I would cut the trees to simplify the building of each wall. Our original cabin set against a sort of natural bluff which rises to become the hill climbing behind our house. I spent nights and weekends through the winter of 1988 making space to build. I worked with a pick, then jack hammer, but eventually hired a back hoe for a day to scoop out enough of the bluff to make ample room for the addition.

This left a section of bluff about seven feet high surrounding two sides of the planned addition. I decided to cut the logs for the upper section of the side and back wall in the woods above the bluff. This would make it easier to slide these logs into place from above. The logs for the lower half and front of the addition would come from the woods on the hill opposite the house.

As I scouted trees I took the four kids along so they could help pick out the logs. Each had a chance to select and mark the trees with paint which would be used in their rooms. Later when the trees were safely down and on the ground, they climbed along the logs, and tried to imagine the rooms they would become.

Over the years here I've seen how quickly a fallen tree can rot when left in contact with the ground, and any moisture can accelerate the process. Since I knew I would have to fell the trees months before I could use them, I decided it would be best to cut the trees when their sap was down and they held the least amount of moisture. By my calculations, I figured this would be in the weeks leading up to the first day of winter. Once the days start getting longer, the forest slowly begins rolling back to life and trees start drawing water up from the ground in preparation for a new years growth.

So I cut the trees in the early December, removed their tops and cut away any branches to give me logs roughly 2 feet longer than the length I would need for the different walls.

The fresh cut trees, even with most of their moisture down, were extremely heavy and more than I could safely move. Some of those green logs must easily have weighed more than 1000 pounds. I planned to begin hewing the logs into shape for walls the following spring, after they dried to a more manageable weight. To protect against decay during those winter months, I propped each log up off the ground at least 12 inches at the ends and in the middle on rocks and small sections of log cut from the tops. It was a chore lifting the logs those few inches!

That winter there were logs stacked everywhere throughout our woods. I checked on each every week or so to make sure it remained safely off the ground, beyond reach of damage. By spring the logs had dried enough that I could easily lift each end into the air. Also, the shrinking which had taken place during those months made it easy to peel away broad rounded sheets of the bark.

I had never worked with a broadaxe or adze before, so hewing the logs proved to be quite a learning experience. I started on May 19th, working in the shade of the woods in full bloom, and continued hewing the logs on and off for almost a year before I was through. At the time I was working in Nashville, a 60 mile commute each way, so I could only focus on the project on weekends, and the occasional day off. Later that summer, when I quit my job to work as a freelance writer I could devote much more time to my project.

Before I made the first swing of my broadaxe though, I had to improvise my own system for hewing so all logs would be roughly the same thickness, and the squared off faces would run nearly parallel to each other.

After much thought and a careful review of the Foxfire book and other cabin books I came up with a pretty simple solution for marking my logs in preparation for "scoring." When you score the logs you make a series of cuts which make it easier for you to remove the unwanted part of the log with your adze or broadaxe. My method required a level, a chalk line, plenty of chalk, two pieces of board six inches and about 1 foot longs, and some nails.

First step was to roll the log so the side I wanted to work on was facing up. Then I would lift and prop each end of the log off the ground. At first, I propped the logs on concrete blocks, but after striking them a few times with my broadaxe I tired of resharpening it and used small sections of log, and wooden wedges, to prop them up.

Once the logs were propped this way, I placed a six inch scrap of board across the butt end of the log , rotated it until I knew it was level with my level, then nailed it in place. Then I repeated this process at the other end of the log. Once I had both boards level and in place, I stretched my chalk line from corner to corner of the boards along full length of the log, and snapped the chalk line until I had a discernible blue chalk line on the face of the log. Again, I repeated this process from corner to corner on the other edge of the boards. Now I had two lines which to serve as guides on how deep I should score the log

I did use a bow saw and buck saw for some of these score marks, but quickly resorted to my chain saw in order to save time. With the log face up and the blue lines of chalk line as guides I cut into the log about every 8 to 10 inches until the saw just touched the lines. On average, these score marks were four to six inches deep. Once they were made, end to end, the log was ready for hewing.

I started working with the broadaxe, but after a few logs, the constant hunkering down and swing of the ax started taking a toll on my back. I ended up doing most of the work with the adze.

When working with the broadaxe, you flip the log 90 degrees after cutting the hash marks, so they are perpendicular to the ground, and then begin the work. There's an art to it using a broadaxe I never fully mastered, but I did l an adequate job. A broadaxe should have a handle which bends away from the head, so you can freely swing the ax into the log without risk of striking the log with your hands. Still, I managed a few skinned knuckles with well intended but misplaced swings.

It's a heavy tool and you have to carefully control each strike so you hit right on the chalk mark. Do it right, and chunks will fly from the log, score to score. If you miss the mark, though, especially if you cut too deep into the log, you've created a lot of additional work.

I swung the broadaxe ahead of me as I worked my way down the log, but I've seen others who started at one end, then back their way along the log. I guess it's a matter of preference as the results appear to be the same.

What I did learn, though, is that hewing a log is not like chopping down a tree. You've got to be more deliberate and methodical in how you swing that ax. It will give you a clean smooth face, but I found swinging my 13 pound broadaxe for hours on end to be some of the toughest work I ever faced. I also found that, as a novice, no matter how careful I was with the broadaxe, I still had to go back over each side with the adze, and clean it up.

Unlike the broadaxe, when you work with the adze you want your log to be face up, just like when you make the score marks. Eventually, I decided to just do all my hewing with the adze, and did a much better job. I can look at the logs on my house now and pick out the ones which were done with the broadaxe, and which were done with the adze by the smoothness of the surface.

I did discover, and rather quickly, that for a klutz like myself, using an adze can be much more dangerous, if you're not adequately protected. I wasn't...at first.

When you work with the broadaxe you are always leaning forward, striking the wood slightly ahead of you. The chunks of wood which fly tend to fall down and away from the log.

Working the adze is entirely different. You're looking down at the surface you are working, straddling the log between your legs. You also swing the adze toward you in a downward curve with your chalk line as your guide of where to hit. It takes practice to control those strikes so they hit just right, neither too high nor too deep.

I must have been working the adze on my third or fourth log when a missed swing allowed the blade glance off the log and into my shin.

Ow! A spot of read soon spread to a crimson patch on my jeans.

It bled for a while, but didn't leave much of a scar. I regard it as a personal reminder of the need to think about safety first, whenever I work.

I kept working with the throbbing wound through that afternoon. But from that day forward I wore a pair of chaps, fashioned from leather and Styrofoam, to cover my shins whenever I worked with the adze. The spared me more than one additional scar. Shin guard worn by soccer players or baseball catchers would also do the trick.

Still, he longer I worked with the adze the more I learned about the value of protective gear. As you work with he adze the chunks of wood can fly up from the log with a lot of force. After one such chunk struck me squarely in the eye, leaving a shiner, I made it a habit to *always* wear safety glasses.

At that point I thought I had it all covered, as far as protecting myself goes. Then one afternoon a wayward chunk of wood flew straight up and caught me in the family jewels. It had enough kick in the crotch to knock me on my back, where I lay a few minutes pondering the clouds passing overhead. As soon I as I was back on my feet I headed into town and purchased an athletic supporter and cup, and wore them whenever I picked up that adze. From then on, I was fully protected against any more incidents.

Over time, as my skills improved, I found the process of hewing to be what I call "philosopher's work." As long as you plodded along, and didn't rush, and were cautious, you could enjoy the process and progress and lose yourself in thought. I had time to think about everything.

The work moved at enough of a clip that I always had the sense I was accomplishing something, but it could take anywhere from an hour to several hours to do one log. A knotty log, especially, presented a lot of challenges which took time; the wood in some logs just seemed harder, or tougher, and slowed the process down.

I enjoyed the process, and felt fine as long as I kept swinging that adze. By the end of the day though, when I finally stopped, my back was extremely sore. If I sat down I had trouble getting back out of the chair; any morning after a full day of hewing I had real trouble turning over to get out of bed.

But the body is like a cold engine I guess. No matter how sore or stiff my back was in the mornings, once I was working for a half hour or so I was back in the groove, and the work moved along.

I worked hewing the logs on and off throughout the summer as I could. At one point, my brother-in-law John volunteered his help for a few days while he and his family were here vacationing. By mid-summer I had more than half the logs ready to go for starting the log addition. I optimistically deluded myself into believing I could complete the work by Christmas when I started framing the floor late in August.

With my limited skills and experience I recognized it would be too much trouble to try and set my floor joists into the logs. So, I attached and framed the floor for the addition as if I were building a deck onto the old house. The approach, though certainly not conventional, gave me a couple of advantages which I adapted for subsequent log cabin projects.

For one, it was much easier and faster than trying to set the floor rafters into the logs. With the floor framed in, and using it as a guide, I could start building my log walls around it to the right dimensions.

Hold on a second, I'm getting a little ahead of myself.

There's about 1/10 of an acre in total level land on our farm, and it's scattered in several places. By late summer the logs I had finished hewing were scattered about the ridge top as well as along the steep hillsides, wherever I had worked on them.

I didn't have a tractor, and I was my own mule. My next challenge was figuring out to move the logs out of the woods and to the work site.

I recalled seeing a TV show once which offered theories about how the Egyptians may have been able to build the pyramids without the benefits of modern machinery. One segment showed the Egyptians rolling each large block of stone on a series of log wheels, which they placed one before the other as each stone slowly rolled forward.

It was a solution I could adapt to my challenge in moving these logs by myself. I cut several pieces of logs, about 4 feet long and 12 inches in diameter, for wheels. By placing them under the logs, I was able to push each through the woods, for 10 or 20 feet before I had to reposition the wheels. Pushing along, then moving back and forth to move the wheels, allowed me to roll the logs to the edge of the woods where I could hook them to our family station wagon, the nearest thing I had to a truck.

I employed this "wheel" method when moving the logs along semi level pockets of the hillside, and again as I positioned them in place at the work site so I could build the walls.

But things eventually got easier, at least as far as moving the log out of the woods goes. One evening I was driving past a roadside landfill when some shiny bundles caught my eye. One of the utility companies had apparently dumped several bales of the wire cable they use between utility poles. I later learned they can't reuse the cable once it's down, and for good reason.

Apparently the slightest kink in the cable can cause it to break, as I discovered once when using it to control the fall of a tree I cut near a roadway. The cable snapped just as the tree started to fall, and the tree twisted across the roadway. Traffic stopped for as long as it took me to cut and clear it out of the way.

But these bundles, each containing between 100 and 150 feet of cable, seemed just what I needed for pulling logs from inaccessible areas of the hillside. I retrieved two of the half inch bundles, and another of 3/8 inch cable.

With these, one end tethered to the log and the other hooked to the bumper of my 1977 Pontiac Safari station wagon, I pulled many logs up the hillsides, and out of the woods to where I needed. That old Pontiac was a boat: it had had a 455 engine with a four barrel carburetor so there was plenty of power. But it didn't have the traction needed to pull and claw through he woods, especially when the ground was at all damp. I spent more than one afternoon with a come-along and shovel, digging the wagon out of a ditch dug by its spinning wheels, then pulling it to firmer ground. This all slowed my progress, but one by one I moved the logs.

At the building site, I could only drag the logs to within about 30 feet of the house. The rest of the way I had to resort to the log wheels which had proven so efficient for moving through the woods.

With the floor now framed in, and the logs on site, I finally got down to setting the first courses of logs in place. Where I was building the new addition onto the old house I had sunk upright pieces of angle iron into holes filled with concrete and then bolted the iron into the walls of the old house. Before I set the iron in place, I had taken each piece to a machine shop to have holes drilled into both sides every four inches. I could bolt each piece of iron to the house, and then to the logs as I raised the walls. For that, I used five inch lag bolts.

When building this addition, I had to lift each log in place, mark it for notching against the lower log, take it down again, cut the notch, and lift it back up before I could finally set it in place. Sometimes it took several tries at trimming the notch before it set right. That mean lifting and re-lifting, with minor trim cuts in between, to get it right. It didn't take too much of this to realize there must be an easier way.

There are other types of notches you can use, but I prefer the traditional dovetail notch. It takes longer, but looks the nicest, and once you've notched your logs together they are secure: it's physically impossible for a log held in place with a dovetail notch to roll out of the wall.

For the remaining notches, I used the dovetail notches from our original cabin to create a pattern. I traced the shape of the notch onto a scrap of pine board, then cut the pattern. This gave me a template for marking and cutting all notches. (The template now hangs on wall of our family room; curious visitors have a tough time guessing what it is.)

It took a while to get the hang of the dovetail notch. In cutting this notch, you work the top and bottom of each log at opposing angles, so when they are placed one on top of the other they lock in place. I came close but never fully succeeded in making a notch as tight as those on our original cabin. Still, you can track the improvement of my skills as a "notcher" as you look from the bottom to top of each corner.

I wanted the addition to closely match the old cabin, but I also wanted to make it clear it wasn't built at the same time. For the mark of distinction, I decided to use an overhanging dovetail notch. The notches on our old cabin are cut flush with the wall; for our addition I cut them so they would overhang about six inches from the wall giving the addition a slightly different look.

By the time I was raising the walls the logs had shed more than half their original weight. I was confident lifting them into place, one end at a time, until about the fourth or fifth course, when the cabin was shoulder high. I always wore a helmet, but was afraid of the harm a log could inflict if it fell just a few feet from above my head.

Now the challenge became devising a way to raise the logs up high enough to build the remainder of each wall. From what I've read, the pioneers used several methods. One was to roll the logs up other logs which served as a ramp. This required several men to push the logs in place, or a mule, ox or horse to pull the log up the ramp. I'd also seen cabins where they would place pegs in the outside wall for a primitive type of scaffolding. then lift the logs end by end, peg to peg, back and forth up the wall.

Even I thought that looked like too much work! I had to find some other solution.

After much thought I decided I could build some giant tripods, and use these to hoist the remainder of the logs into place. For the tripods, I drew on skills and experience gained as a Boy Scout. I got out my old Scout handbook for a refresher course in lashing, then built tripods using 16 foot locust saplings, about four to six inches in diameter. I chose locust because it's one of the strongest woods in the forest here, and doesn't rot in the weather. Also, it's easy to find your standing dead locust trees with a relatively straight 16 foot run.

To actually raise the logs, I bought two sets of block and tackle, one new, the other an old one from our neighbor Quinton.

By the time I was ready for the tripods I had not nailed any board to the floor rafters. I planned to use plywood for a sub floor and was worried it would warp if left exposed to the wind, rain and snow while I worked on the cabin. Instead, I just placed planks of wood over the rafters to make a walkway.

Since the cabin was about two feet off the ground I figured it wouldn't be too difficult to maneuver the tripods into place. I did have to lift the tripod legs off the ground three feet to get them past the floor rafters, and then move them back again. And since I had only bothered to build two tripods, I used the same tripod for the side and front wall(which meant I had to move them around for each log).

Raising the upper logs required a pair of tripods, block and tackle and long poles supported on the short bluff.

As I review the process now, I realize I got pretty good over the course of this project at figuring out ways to give myself more work.

Anyway, I quickly discovered this system worked, but it could also be very dangerous. First thing I learned was the need to brace the legs of the tripod to each other at the bottom, as they could slip under the weight of a large sections of log. Simple solution: I tacked scraps of board from leg to leg, then knocked them loose when it came time to move the tripod.

The other danger was in that old block and tackle bought from my neighbor. It had a tendency to slip, or give, when it was supposedly locked in place, allowing a raised log to slip several feet before it locked tight again. I made it a habit to tie the pull cord to the floor once a log was in the air so there was no chance of slipping. And I never let go of that rope while raising a log.

Needless to say, the building of the cabin proved much slower than I'd expected. As I said, I had honestly believed we'd be in the addition by Christmas. But in early December 1989, the first ice storm of the winter hit, and my walls were only window high.

That proved to be a cold winter for us, too; I set my tools aside and awaited the thaw. At least there was plenty of kindling for our wood stove from all the scraps I'd knocked off while hewing out the logs.

So we spent one more crowded winter in our five room house, making plans for the rooms, the indoor bathroom and shower, which would be ours within months. Warmer weather took a long time coming, but I was able to put another course of logs up in February-March, and fully resume work on the addition the first week of spring.

First thing I did was build the box frames for the windows. Before I could nail them in place , though, I had to chisel flat spots in each log so they could set level. That done, I nailed the boxes to the logs, and braced each to the floor with a scrap of wood to keep it plumb. Window boxes in place, I had a guide for the shorter pieces of logs I could now use to complete most of the walls.

There was still a lot of work moving the tripods and cutting the notches for some of the longer sections, but it was lot easier than working with a full length long. And, the higher the walls climbed the easier it was to get a real sense that I was building a log cabin. We were inspired!

That spring I also had a regular audience for my work. Our local gas company was running a new line down the old highway by our house, and there were crews around all the time, watching me work in their idle moments, pumping me with questions, encouraging the progress.

Now that I was working with smaller pieces of logs, I also felt safer about having Liz and the kids around while I worked. They were pretty good at helping me misplace my tools, too, and they had a lot of fun running the boards which served as a walkway until I was ready to put down the subfloor.

As I've mentioned before, I knew the cabin was going to take some time, and I had intentionally postponed putting down the plywood subfloor until I had the roof overhead. I had seen how sitting water can damage such a plywood floor. As an interim solution I had simply just laid planks over the rafters without nailing them in place to serve as a walkway. This proved a most foolish and regrettable mistake: one time Liz was helping me she slipped from the walkway and fell onto one of the floor rafters at full force. She was sore for weeks.

Anyway, the more it looked like a cabin was taking shape the more excited we all got about it. Working as we could, in the evenings and weekends, when there wasn't some other chore to tend to, the cabin walls slowly climbed higher through April, then May. Finally, as the sun was preparing to slip into our valley on May 18, 1990 I hoisted the last of the full length notched logs into the air. But just as I prepared to set it in place, that old block and tackle gave way and the log went crashing to the ground. It missed me, by inches.

I caught my breath, tried again more cautiously, and finally rolled the log into place. The sun was already set when I climbed the log wall at the corner and spiked it into place. Liz and the kids were in the yard to watch, and cheered.

It took a full year, exactly, from when I started hewing the first log, until I set the last one in place on the addition. it was just a shell, but the cabin walls were up. Granted, had I worked at it full time the project would have moved much quicker, but the toughest part of it, the hardest work I've ever done, was over. I'd proved to myself that I , or anyone, could put together a cabin with no real skills but the willingness to work and the determination to see the project through,

I stepped into the road with Liz to look it over, talk about the experience, and ponder the possibilities of the new rooms.

The next day, my 35th birthday, I took a break from the work, but resumed the following morning, eager now to push the project to completion.

With all the logs in place I dismantled my tripods, undid the lashings and cut the poles up for next winter's firewood. Now that they were out of the way, we could lay down the subfloor. For this, Liz and I had an eager crew in all four kids. Each got a hammer, and they were content to bang away at the floorboards, with or without nails. They were just thrilled at the opportunity to help.

Once that floor was down, Liz and I framed the inside walls from 2x4s , and set them in place to divide the one large room into several smaller ones.

Before we could build the roof, I had to take steps to make sure the top logs at the front and back, which would serve as the plates for the ceiling and roof rafters, were level with each other. That meant tedious work with a hammer, chisel, plain and level. After several hours of chipping and shaving away, I stretched a line from front to back wall and the line level told me that part of the work was done.

Then, we moved on to the ceiling and roof. Back in the 1960s some speculators had come through this area looking for oil and had actually built a test well on my neighbor Quinton's farm, across the road. When the drillers gave up they left him all the lumber they'd used to build a deck and scaffolding around the test well. He took it down and stored it in his barn until we bought the 2x6x12s and 16s from him for $2 each. They served as the ceiling rafters for our addition.

A Hand-Hewn Addition, The Old Fashioned Way

Plenty of help for putting down the floor then dividing the large log room into several rooms, framing the walls with 2x4s.

We realized I didn't know enough about framing to even attempt to build the roof myself, so I hired friends Bob and Graham, to help . Actually, Graham and I helped Bob, who is an accomplished carpenter and cabinet maker. He framed the roof in a day so we could have a cathedral ceiling in our family room. Later, Graham also helped me space and nail down one inch boards for the lathe, and then attach the tin for the roof.

The one day Graham couldn't help me, I decided to go ahead and work on my own, putting down the long sheets of tin. While carrying one 12 foot sheet across the roof, I slipped through a gap in the lathe, lost my balance and the tin sheet I'd been holding. As it dropped, it glanced along my pinkie just about slicing it off.

My leather gloves were neatly folded in my back pocket. That cut required several stitches to secure my finger. It gave me a few days rest from roofing to ponder this latest safety lesson. It was July, and very hot anyway.

One of the nice features of the addition was placement of a skylight in the roof directly above what we call our family room. It's a great effect and really brightens up the space, but the damn thing is a source of persistent leaks , which I've never been able to completely eliminate, though I've spent hours on the roof with caulk and tar. I've heard similar stories from other amateurs who have installed skylights, so a word to the wise…

Once that roof was overhead finishing up our log addition became more like any other building project. I do want to point out this though: running the electricity turned out to be a major hassle. I had not thought ahead and planned how to run the wires from room to room through the logs when putting up the walls.

Liz did most of the actual wiring of each room—she didn't entirely trust my attention to detail where safety is concerned—while I handled the grunt work of pulling the cable over, above and through the walls. Ultimately, we decided the easiest solution was to run most of the wiring to each room from above, avoiding as much as possible the need to drill or cut through the logs. In the gaps between the logs we ran all the wiring through conduit. Where we could, we also hid the wiring behind trim around windows and the door.

Only after the cabin addition was entirely wired, did we turn to chinking the gaps between logs. For this we used the same methods, more or less, outlined in Part I. Once thing I did find different about this cabin, though, was that the logs continued to shrink for several years after they were put up. Even though I had allowed them to dry in the woods before building, they still shrank away from the chinking in those first few years. Liz got pretty handy with the caulk gun filling in the cracks to keep out the air leaks.

The kids didn't wait for chinking though; they all moved into their rooms as soon as there was a roof overhead, and windows in place. It would be the fall before we finished with the drywall in their rooms.

The week before Thanksgiving we also hired our friend Chip, an experienced mason, to build a chimney and brick mantle for our wood stove in a corner of the family room. I helped as best I could, but it was his expertise which made such a nice feature for the room.

At the base we put a family "time capsule." Just a couple of glass jars with odds and ends from around the house: a spare TV remote, packets of seeds, family photos, tiny toys and whatever else made sense. We also set arrowheads and stone tools the kids and I had found in the area into the mortar between the bricks to add a local, historical touch to the brick mantle.

By Christmas all the work was finally done and we were all fully moved into our new space. As part of our celebration we went out and cut down a 12 foot cedar to put up as a our Christmas tree in the family room. Our place looked picture perfect, with the tall tree decorated against the log walls .

The kids all had their rooms, Liz had a laundry room and we had a cozy new family room with a beautiful chimney. But I think the biggest thrill for everyone came with the move of our bathroom facilities indoors, and with a bathtub and shower!

Liz and I can't help but smile when we recall the sight of all four of the kids gathered around the toilet right after we finished installing it, taking turns at pulling the handle, straining to watch the water disappear down the bowl.

Now that they are older they put on airs about city sophistication, but we know better, and have the memories to prove it.

Our home today. At center is the restored original one room cabin, which we built around and above to make room for our family, using hand-hewn logs for the large addition at the right, and modern framing methods for the kitchen at left, and room at top.

2

Methods for Building A Hand Hewn Cabin

In these pages, I discuss the methods used for a hand-hewn log cabin. You should also refer to the sections on tools and modern construction methods in the back of the book, the chinking section in Part I, Chapter 2, and section on framing the floor "Prepping Your Site" in Part III, Chapter 2.

When we planned the addition for our cabin we knew precious little about building, or the amount of work we were assuming. We found the most important tool for the project to be our determination to see it through. For us, building with hand-hewn logs proved a slow process, shaped in part by availability of funds. Having the right tools and materials on hand will greatly reduce the amount of time involved, and some of the labor. But, no matter how you approach it building a log cabin is a labor-intensive undertaking.

Many of the considerations discussed here apply to any log cabin, as well as a hand-hewn log cabin. I am fully aware of my limited skills experience, and knowledge of many aspects of carpentry. I do not want to mislead you and overstep those limits. The methods outlined throughout this book are drawn from our experiences. When I suggest you refer to additional sources it's because I think you need more advice than I can comfortably provide. There was a lot of improvisation in my projects, and I am sure there will be in yours. When you have to turn outside for help, be it a book or a builder, always use the best resources available to you.

In composing this section I assume you have a basic understanding of tools and building concepts like "plumb," "level" and "square." If not, as I was, I am confident you will quickly master these by spending a little time reading a general book on carpentry or construction.

All of the information which follows is presented in the logical order in which you can expect to face these challenges.

Selecting The Site

What matters to you?

A panoramic view…or the seclusion of a retreat hidden in the woods?

Cool summers…or warmth and light in winter?

Your cabin will be a great source of pride, and the home you select for it will shape how much you enjoy the time you spend there. Before you consider any potential site, you need to know what you want in sources of water and energy.

Some may want hookups with the local water, gas and power companies; others might be willing to haul water to the cabin equipped with only a wood stove for heating and cooking. Deciding these priorities will help narrow your selection to the sites which satisfy your needs today and tomorrow. A reliable water source, year round, should be one of the guiding considerations when comparing sites.

If the land will also be the source of the logs you'll want to make sure there is an ample supply of usable trees in the immediate area, or they can be easily transported to the site.

Prolonged exposure to moisture is a log cabin's worst enemy. It will invite and accelerate decay of the logs, making them susceptible to a range of insect infestation. Therefore, the first thing you want in a cabin site is an area which is well drained, preferably slightly elevated above the surrounding terrain. That way rain or water can easily drain away from the structure.

Ideally, the site should also be accessible on all sides when building the cabin. You can build your cabin near a bluff, as I did, but it creates more work when moving and raising the logs. When considering any site, remember also you will need space nearby for storing building materials, including the logs, and the many tools the project requires.

It's also important the site be accessible by an existing road, or you can easily cut a road to it. A friend carried all his building materials to the site of his retreat in the woods, piece by piece. Most aren't up for that kind of physical challenge.

Next, consider the direction your cabin will face when evaluating a site. A view to the east invites the suns first rays, while looking west opens to beautiful sunsets. In the U.S. the weather, and wind typically come out of the west pushing east. Northern exposures, while cool in summer, can be extremely cold in winter.

Every cabin should have a porch, and the direction it faces can help shape your enjoyment from it. Ours, which looks to the northwest, always has a comfortable breeze in summer, but it catches the wintry gusts of December and January. Conversely, a deep porch facing south provides a shady haven on the hottest days, yet will be warmed whenever the sun rides low through winter.

In the old days, when space was tight, the porch might have served as the summer bedroom on hot nights, a gathering place at the end of the day, and an informal place to entertain friend and enjoy some homespun music. It can and should be just as vital a center of activity in any cabin built today. And if you can't have a wrap around porch, how about a front porch, and a deck out back?

If you plan on using solar energy you'll need to situate the building so one side of the roof faces south for installation of solar panels in the roof, unless they will be self-standing and separate from the building. For wind power, you should know the direction of prevailing winds so you'll know where to best harness them.

THE LOGS

Once you know where and how your cabin will sit, start thinking about the building. Carefully plan the size of the cabin, both in the length and width of the room and the height of the walls. The number of logs required depends on the diameter of the logs you use. For my cabin, working with logs that averaged from 10 to 14 inches in diameter, I used an average of 5 logs for every four feet of vertical height.

As far as the length and width of the building, and length of the logs, decide first how large you want your living space to be. Otherwise, if you plan your space by the outside dimensions of the cabin you'll end up with less home than you think.

Where I live in Tennessee, the standard one room cabin measured 16 by 18 feet. If you want a cabin with an 16x18 usable area inside, and will hew the logs to the standard thickness of six inches, you'll need logs at least 19 feet long for the long wall, and 17 feet long for the short wall.

I recommend, though, that you always start with logs at least two feet longer that you actually intend to use, and trim them down later. So, you would need 20 foot logs for an 18 foot wall, or 28 foot logs for a 26 foot wall.

When you have your dimensions and the height of the cabin you can estimate the number of logs you'll need. We'll use that standard 16x18 foot cabin, with an interior ceiling height of 8 feet, as an example. For the purpose of this example, figure you'll need five logs for every four feet of height.(Remember the diameter of the logs has a direct bearing on the total number of logs required for each wall.)

For each side we're going to need 10 logs, plus 1 course of logs surrounding the floor sills and rafters before we begin building the walls.

Based on these assumptions you'd need a total of 11 logs for each wall, or a total of 44: 22 logs at least 18 feet long, and 22 at least 20 feet long.

That's a lot of trees!

I mentioned earlier with the traditional method, the entire cabin walls would be raised before they'd cut and frame the doors and windows. If you want to reduce the number of trees you need to cut, you can do as I did and plan the size and location of each door and window before hand. Planning ahead in this way, you can reduce the number of full length logs required for the project by as much as 1/3, even more if you add a lot of windows.

With this approach, though, you need to determine the approximate size of each door and window first so you can plan the size of the opening for each. Then, draw your wall plans with these boxes in place, and you can estimate what you'll need in total log length for each course or layer of logs of the cabin.

No matter what your tally, plan for a few more trees, to be certain you'll have enough logs. Cut 10 percent more than your best guestimate and you should be covered.

SELECTING YOUR TREES

When you have an idea how many trees you need, the work turns to locating them. The first thing you need to know is what types of trees you should use.

The pioneers built their cabins of whatever trees were available on site, and some of those cabins have lasted hundreds of years. Later, as subsequent generations built cabins of their own, they were more selective in the trees they used, for a variety of reasons.

The ideal log for a cabin is relatively light and straight, easily worked but durable. Where I live in Tennessee yellow poplar and chestnut became the preferred woods, and sometimes cedar. Poplar still grows in abundance here.

In other parts of the country species of pines or firs may be most commonly used for cabin construction. The ideal tree is tall, decay resistant, and without much taper over the length of a log.

If you don't know what type of local wood to use, ask around. Some old timers may still be able to tell you what trees were preferred, even help identify them in the forest. Visit surviving log cabins in the area and find out what types of trees were used in their construction. Apart from native American chestnut, odds are those are still growing throughout the area.

You could also talk to a logger. He knows as much as anyone about the qualities of different types of woods, and where to find the various species growing. In fact, if you don't want to cut and haul the logs yourself, your best source will be a logger. One can have all the logs cut to your specs, and delivered to your site.

Most of you, though, will cut the trees yourself. You'll be headed into he woods in search of perfect logs. If you don't know your trees well enough to identify them by the bark, I'd suggest you head out with a tree guide and mark them during the growing period, when its easiest to identify them by their leaves.

Finding trees of the right length may not be as easy as you think. You need a long straight run, with a log that won't taper much in diameter from butt end to butt end.

There's some things to avoid when selecting your trees, too. Many branches in the log section means there will be a lot of knots. Knots can slow the process down when hewing the log. Later, the area surrounding the knot may also be especially susceptible to decay or water damage.

Any signs of decay in upper branches of a standing tree, especially hollow openings into the trunk, suggest the entire log may be rotten. That's a dangerous tree to cut.

Also try and avoid any tree which twisted around its trunk as it grew, even slightly. This happens when the tree has been under a lot of stress from prevailing winds, or compensated for some proclivity to lean in one direction, as it grew. Logs with these turns can also prove especially difficult to work with an axe.

When you think you've got a tree with all the right features, walk around it at its base and look up to its crown from all sides. Sometimes this will reveal a bend or turn in the tree which you might otherwise miss.

If a tree meets all your criteria, go ahead and mark it for cutting, or prepare to bring it down.

WHEN TO CUT?

Over the years I've heard different theories on when and how to cut your trees for durable logs. One theory holds you should "ring" your trees during the growing season by removing a band of bark about a foot wide all around the base of the tree. This will kill the standing tree, and within a few months the water will drain out of the log portion without significant "checks" or cracks in the log.

I've also heard you should cut down trees as they are beginning to grow in the spring, and leave their tops attached to draw all moisture out of the trunk before you cut the sections for your logs.

Another theory, and the one I followed, is that you should cut your trees during their dormant season, when there's a minimal amount of moisture in the sections of the tree above ground.

I've never heard anyone recommend you cut trees for a cabin in the thick of the growing season. Nevertheless, I would venture a guess that many old log cabins—and some have survived centuries—were put up using green logs cut in the height of summer!

Yes, they've lasted well…but it must have been some challenge raising those heavy, fresh cut logs.

Whatever season you choose, plan to give your logs some time to dry out before you begin working them. And I'd also recommend you don't wait more than six months between the time you ring or cut the tree and you actually start working the logs. If you wait much longer you could have problems with insects and decay, or end up with seasoned wood, difficult to work.

BRINGING THEM DOWN

Cutting trees is dangerous work. If you don't know what you're doing you could get seriously injured, easily! Every experienced wood cutter knows any tree can break or fall where you don't want it to. So cutting a tree is not something you can just head out to do with a little study. If you don't have experience, or have any doubts about your ability, you should have someone who knows what they are doing show you how to control the fall of a tree. Better yet, have them cut the trees for you.

By all means don't rely on what you read here alone to guide you through this process. It is your responsibility to learn and follow the steps involved in safely bringing down a tree. Even experienced woodcutters would do well to refresh themselves on the proper way to fell a tree. Any chain saw manual and most forestry or landscaping books include diagrams explaining how it's properly done. Consult them before you start cutting your trees.

And don't forget your safety equipment: hard hat, glasses, ear plugs and gloves.

Never cut your trees on a windy day as you may have no control over where and how they fall.

Most trees lean in one direction as they grow, even slightly. The first thing you should do is look up the tree from around the base of its trunk to see where it's naturally leaning. A skilled, practiced logger can make a tree fall almost anywhere he wants, but the easiest way to bring down a tree is to cut it so it will fall in the direction it already wants to go.

When you look up into its crown you also want to make sure the tree is healthy. Rotten limbs, may indicate hidden damage inside which can cause the tree to suddenly break while being cut. Also make sure there are no heavy vines or entwined branches which may prevent the tree from falling, or bring down other trees or sections of them as the tree falls. These can be especially dangerous and cause something to fall around the base of the tree.

Once you determine that the tree can be safely cut, and where it is leaning , trace a path through the woods where you expect the tree will fall. Smaller trees in its path can cause the log to break, or the tree to buck back at its base. Clear its path of anything which could impede its fall.

Before you begin cutting a tree you've got to give yourself a clear area to work in, and several possible escape routes in case the tree suddenly breaks or starts falling where you didn't want it to go. Clear your escape routes of anything which could trip you if you needed to run away suddenly. Make sure they allow you to quickly get a safe distance from the base of the tree

You begin felling a tree by cutting a notch into the trunk on the side in which you want it to fall. It's the same whether you cut with an ax or a chainsaw. Remove a wedge or triangle shaped area. First cut down into the tree at an angle, then cut up so you remove a wedge which penetrates about 1/3 into the diameter of the tree.

As you cut the wedge continually check to make sure there's no sign of the tree suddenly breaking or leaning away under its weight. If so stop cutting and consider your options. You may need to use chains and a cable hoist to safely bring the tree the rest of the way down.

When the tree poses no danger, once the wedge is removed, your next step is to start cutting into the tree from the side opposite the wedge. Start cutting straight into the tree about least two inches above the center of the wedge. When done right the tree will start leaning into the wedge and soon fall under its weight, with the area between the wedge and this new cut serving as a pivot to guide the tree down.

All the time you are cutting into the back side of the tree, opposite the wedge, regularly look and listen for signs that it is leaning or beginning to fall. As soon as the tree starts its way down, remove your saw, turn it off turn off and hurry away using one of your predetermined escape routes.

Move a safe enough distance away so there is no danger you could be struck if the tree suddenly lurches in a different direction or bucks back at its base before hitting the ground.

CUTTING THE LOG

With the tree on the ground it's time to cut or buck your log(s) from the trunk and stack them, so they can dry without risk of rotting. Begin by removing the top of the tree just below where it begins to branch to create the crown. Note: if you've cut a tree with its leaves during the growing season, you may want to wait a few weeks before cutting away the crown so the leaves can draw remaining moisture out of the log section.

For safety's sake, make sure you can see what you are cutting. Always clear away smaller limbs and other trees which could catch or impede your saw before you cut the log. If the tree is under pressure, make a cut about 1/3 of the diameter into the trunk of the side under pressure first. This will relieve the stress, and lessen the chance of your saw being pinched, when you cut on the opposite side to remove the log.

Once the log section is cut free, use your saw or axe to remove any remaining smaller branches growing out of the log section of the tree. You may have to roll the log to get all the branches. A green log is heavy so you'll probably need your cant hook or iron prybar(see tools section) to maneuver the log.

If you're planning on putting up your cabin green you can begin hewing the logs right away(next section). If not, you need to stack each log up off the ground so there's no risk of rotting while it dries out. Cut your sections of log from the trunk, leaving at least an extra foot in length.

Raise these up and stack each up off the ground at least 16 inches. I'd advise you stack the logs on rocks; as stacking them on section of log can invite infestation by insects.

Be sure and check on each log over the next few months, and roll them occasionally so they won't warp.

MOVING THE LOGS

You can leave your logs in the woods to dry, or move them to the building site and stack them there. You can work them in the woods, or hew them as the cabin walls rise. Either way, you need to figure out some way to move your logs, even if it's only into place around the building site.

The challenges you'll face in moving logs depends on how accessible they are and what equipment is available to you. I had a couple of unique challenges, and how I solved them may help you figure out an easier way to move your logs.

Many of my logs were cut in the deep woods. I had to snake them out of the forest before I could even hook a chain to them. To maneuver them, I used a series of log "wheels" cut from sections of the trunk. Each of these sections was about four feet long, and from 12 to 18 inches in diameter. They had to be round and smooth as well, without branches or knots which would impede their ability to roll over the ground. I always had at least three available.

I'd start by placing one of these "wheels" underneath the log, about four feet back from the front end of the log. I'd place the second wheel near the middle, and left the third wheel along the path toward the front of the log.

Once the two wheels were in place, I'd simply push the log forward. As I did the log rolled over these wheels in the direction I pushed, enabling forward progress. I'd push this way until I reached the middle wheel and ran out of pushing room.

If someone was helping me, they could continually place the extra wheels in front of the log as I pushed to keep it moving. Otherwise, as soon as I ran out of pushing room I'd step ahead and placed the section I'd left along the path under the front of the log. Then I'd push again, continually replacing the log wheels as often as needed until the log was where I wanted it.

This method required some muscle but proved an effective way to move the logs through the woods. Later, I adapted the same method, with shorter pieces of logs as wheels, to roll logs into place around my building site before I raised them.

As far as dragging the logs through the woods, it's merely a matter of attaching a chain to them, hooking it to your "mule" and pulling them along. Since you can scrape or damage the end of the log as it is dragged, allow an extra foot or so of log at each end. Later, when your logs are on site or up, you can cut them back to the length you actually need.

I'd recommend at least one logging chain for dragging the logs, and a lighter length of chain around six feet long, which you can loop around the log and hook to the logging chain. Or, you may want to get a set of "skidding thongs" or "dog chains" for attaching your logs. Hooks on the sides of these grab and hold the log as long as pressure is applied on the chain.

Whatever combination of chains you use, make sure you have the extra hooks, removable links, and pins to make it easy to modify your chains as needed for a secure connection.

HEWING THE LOGS

You can hew your logs in the woods where you cut them, green or dry, or haul them out of the woods to your work site and hew them there. In either case, the work involved is the same.

Start by peeling the bark from the log. It's important that you remove the bark, as the space between the bark and log is where many insects will get their start on the wood. When the logs have had a chance to dry a while the bark may peel off in sheets. If the tree is green, or the bark difficult to remove, you may need to use a wide chisel, or a special tool called a spud or bark peeler, to work the bark loose. Simply slide the edge under the bark and strike the tool to loosen the bark. Be careful not to set the edge at too sharp an angle too deep or you will end up gouging into the log.

With the bark removed, you can mark your logs for scoring, the first step in hewing. You'll want your logs elevated slightly off the ground so you can comfortably work them. I recommend you prop them on short sections of log, then secure them in place with wooden wedges. Don't set your logs on stone or concrete blocks, or use iron wedges, as they will damage your tools.

Whether you're going to hew with an axe, broadaxe or adze, the first few steps of the process are the same. Most hand-hewn cabins are built of logs six to eight inches thick, with a flat hewn face showing both inside and out. No matter what the diameter of the log, you still want to work all logs down to the same thickness for building the cabin.

First, roll your log and examine all sides to determine which two sides will give you the straightest run of log. Any twist or turn into or out of the wall will make it difficult to chink. After you've determined the straightest run of the log, roll the log and wedge it so the straightest runs, the sides you will hew, run parallel to the ground.

Cut two scraps 0f board at least one inch thick, a full six inches wide and at least a foot long for guides when marking the log boards (use eight inch wide boards if your walls will be eight inches thick). Set a board across one butt end of the log and hammer in one nail to hold in place so the edge of the board is roughly parallel to the ground.

Then set your level on the edge of the board, and rotate the board until its edge is level. Drive in a couple of nails to secure the board in place. When you repeat this process at the other end of the log, the boards will serve as a guide to ensure consistent thickness of the hewn log.

Then, take a chalkline loaded with plenty of blue chalk. Stretch it across the length of the log from the corner of one board to the corner of the board on the opposite end. Snap your chalkline so you have a dark straight line running the full length of the log. (You may need to hold the line against the log in the middle to get a dark enough line.) When the line can be easily seen, repeat the process, board to board, on the other side of the log.

This will give you a set of parallel lines, running the full length of the top of log, which will guide your cuts in the next step, "scoring" the log.

Scoring is merely a matter of cutting straight across the log to the depth of these lines. Start at one end, and cut down to the chalk line every six to eight inches across the upper surface of the log. When you encounter a knot, or a spot where a branch grew out of the tree, space your score marks much closer, as little as an inch apart.

Once the log is marked with these "scores", also called "hash marks", you're ready to begin hewing or squaring off the face of the log.

Hewing With An Ax

You can hew the log with a regular long handled single or double bladed ax, or use a broadaxe. I'd recommend the broadaxe. Its cutting edge is much wider allowing you to remove more wood with each swing. Another useful feature of a broadaxe is that its handle is curved. This keeps your hands several inches away from the log as you hew, eliminating the chance for skinned knuckles. You should be aware that a broadaxe weighs much more than a regular ax and gives the user quite a workout.

When hewing logs with any type of ax, the face or side of the log you are working needs to be perpendicular to the ground so you can swing down with full force and accurately strike your chalk lines. So, first roll the log 90 degrees so the score marks are perpendicular to the ground.

You can start at either end and work forward or backward along the log. Either way you'll end up with a sore back. I preferred to work forward, swinging the ax ahead of me as I moved down the log.

When hewing, or squaring off the face of the logs, you swing the ax down so the blade strikes the chalk line. As you do this some of the log between the score marks will fall away. Continue to cut down into the log, using the score marks as your guide, until the area you are working is fairly flat and even. Once you have a little practice, you should be able to smooth a small section of the log with each swing.

Repeat this process down the full length of the log. Use your chalk lines to guide your initial cuts, then your score marks. When you're good with the ax your logs will have a straight clean face. I never got quite that good and always went back over the face of the log with an adze to touch it up(See next section).

When you're through hewing one side of the log, flip the log onto the freshly flattened face, snap your chalk line from corner to corner of your end boards, and repeat the process. It can take anywhere from an hour to a couple of hours to complete hewing both sides of a log. It all depends on your energy, the condition of the log and your skill with the ax. The work does get easier with practice.

Working With An Adze

I ended up needing an adze to clean up most of the work I did with the broadaxe. Usually, it was just a matter of knocking away the rough spots I'd missed with the ax. The more I used it though, the more comfortable I felt working with the adze. Eventually I decided to use the adze for hewing the remainder of my logs.

The adze has a wide, flat blade which is slightly curved and perpendicular to its handle. To cut with it, you swing the blade toward you, with the side of the log you are hewing facing up, as it was when you scored it.

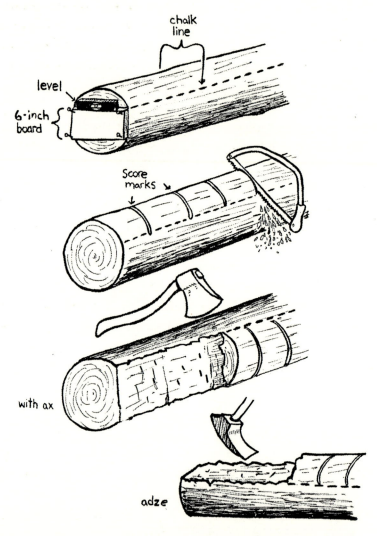

Hewing a log. Tack board equal to wall thickness to end of log and rotate until log is level. Repeat at other end, then stretch chalkline from end to end, marking log. Score log to depth of these parallel lines about every six to eight inches along the log. With ax: roll log so score marks are perpendicular to ground and swing ax down, using chalkline and scores as guide. With adze: leave log so chalkines run parallel to ground and work your way along log face with adze.

Since you are swinging the adze toward you there's always the chance it could glance off the log and strike you in the leg. As I mentioned, this happened to me, leaving a gash in one of my shins. I strongly urge anyone using an adze to wear some sort of lower leg protection, or chaps, along with safety glasses.

When you work with the adze you start at the far end of the log, and gradually back your way to the opposite end, squaring off the surface of the log as you go. It takes some measure of practice to develop the rhythm and control which will allow you to hew the log with precision.

The first cut is always the toughest one to get right with the adze. Use the edge of the board nailed to the log as a guide. Swing the adze down and toward you so it strikes the log squarely, perpendicular to your score marks. When you do it right, the log surface will flake away across several score marks with each cut. Once you've started use the flattened surface of the log and the next score mark to guide your cuts.

After you've worked with it a while, you'll be surprised at how quickly the flakes fly away, and how clean a surface you're left with.

As with the broadaxe, you may need to go back over the surface of the log to clean it up before you flip the log onto its hewed face and work the other side.

NOTCHING THE CORNERS

The are several techniques you can use for "notching" your logs. The notch attaches your walls at the corner, locking the intersecting logs together, strengthening the entire structure.

A variety of notches have been used in building traditional cabins. Some of the most common including the simple lap joint, the saddle notch and the dovetail notch. Our original cabin was held together using dovetail notches at the corners, and that's the notch I decided to use when adding to it. For another building I used the rounded saddle notch(explained in Part III, Chapter 2)

The dovetail notch is probably strongest of these notches. It's intersecting angles locks the logs together, making it impossible to move or roll the log once it is set in place.

To create this notch, the top and bottom of every log is cut at the same angle, usually 30 or 45 degrees, but sloping in opposite directions. For instance, the cut on the upper portion of the log slopes down from the inside to the outside face of the hewed log, for the entire width of the log. The lower part of the notch slopes up at the same angle from the butt end of the log in for a distance slightly longer than the thickness of the log. When cut, this lower notch removes a triangle with a base equal to the thickness of the wall.

Please consult the diagrams for reference.

MAKE A TEMPLATE

You could measure each of these angles and cuts as you go, but you'll find it much easier to work from a template. I created mine using the notches on my old cabin as a model, but you can easily create one of your own with nothing but a protractor, square and scrap of wood.

First, decide on the angle of the notch you want to use. I'll use 30 degrees and six inch log walls as an example.

Start with a piece of 1x6 or 2x6 at least four inches longer than the thickness of the hewn logs you intend to use

Set your square against your board and draw a line representing a 90 degree angle six inches in from the outer edge of the board.

Then take your protractor and set it along the base of the board so you can draw a second line representing a 30 degree angle from the outer edge of the line.

Extend this line from the lower corner at the outer edge of the board until it intersects the line representing the right angle.

This gives you the outline of a triangle.

Cut this triangle from the board and you have a template for your notches.

Now for the notching.

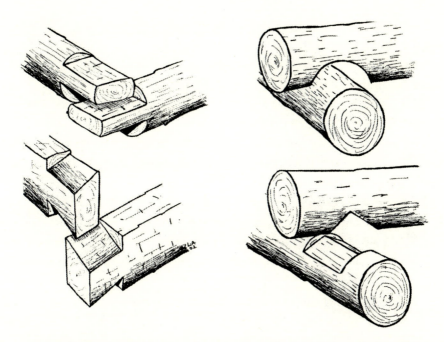

Some types of notches used in log cabin contruction. Top row, from left: simple lap notch, rounded saddle notch. Bottom row, from left: dovetail notch, A-V saddle notch.

For upper notch cuts:

Decide which side of your log will face in, and which will face out. Set your pattern against the log's end so the inner edge of the template, representing the 90 degree angle, is flush with the inside edge of the log.

Trace the slope of the template triangle along the outside of the butt end of the log.

Next set your square against the butt end of the log where this line meets the outside surface of the log.

Draw a line perpendicular to the angle line along the outside face of the log. This line should be slightly longer than the thickness of your log walls.

Next, set your square perpendicular to the start of the line representing the slope, against the inside face of the log.

Trace a draw a line of equal length to the line you drew along the outer face along the upper edge of the inside face of the log.

These two lines will serve as guides for scoring the notch, just as you scored the log before hewing it.

Use a saw to make a series of score marks into the log at an angle so the cut touches both the upper and lower lines you've drawn, along the inner and outer faces of the log.

Next, set the cutting edge of a broad bladed hatchet or chisel against the line representing the slope of the notch across the butt end of the log. Strike your tool with a hammer so it removes a section of wood to the score mark. Continue working this way until you reach the last of your score marks.

Use a chisel to clean up the notch as much as possible.

For lower notch cuts:

After the top notch is cut, flip your log so the hewn faces are parallel to the ground.

Take your template and align the base of the template triangle with the lower edge of the log. The perpendicular line defining the right angle of the template should line up along the inner edge of the upper portion of the notch.

Trace your triangle.

Flip the log over so you can repeat these steps on the opposite face of the log. You should end up with a pair of matching triangles drawn on both sides of the log.

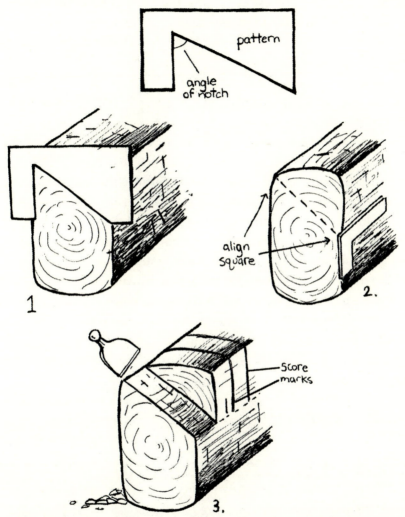

Dovetail notch, upper cut. Make a template like one shown, with angle of notch and triangle as wide at base as thickness of log walls. Align and trace triangle along butt end of log. Align square with top and bottom of diagonal and trace line equal to thickness of wall. Using these lines as guides, score log with saw to angle of notch and remove wood to create the upper cut

You can repeat the process of scoring the log and chiseling out the notch, but I found it easier to make the cut for the lower notch with a saw, working carefully and slowly using the lines drawn from the template as a guide.

It's easier to make the cut if you flip the log back up, so the hewn faces of the log are perpendicular to the ground, and the area to be removed is on top.

First make a straight cut along the lines defining the depth of the notch.

Then starting at the outer corner, slowly cut into the log at the angle drawn, using the lines on front and back to guide your saw. Slowly cut along this line until the cut intersects your other cut.

You should be able to remove a triangle wedge of wood, closely matching the shape of the template. Again you may need to clean up the notch with a chisel before your work is complete.

With these cuts made, the notch is complete, more or less. You may require some adjustments for a better fit after you've set the log in place. Also, if you are working with an especially wide log you may need to cut both parts of the notch deeper into the log to avoid large gaps between logs in the wall.

After setting a few logs in place you'll understand how you can make these corrections to control the size of the gaps in the wall.

PLANNING FOR ELECTRICAL AND PLUMBING

(For more information see the Adapting the New appendix at the back of the book)

If you want modern amenities, take time to figure out how you will run your electrical wiring, plumbing pipes or heat pipes and vents through the cabin before you start putting up the walls. If you don't plan ahead you'll leave yourself open to additional work, and frustration, later on. It's much more difficult to retrofit a log cabin with pipes or wiring than it is in a modern framed house.

Methods for Building A Hand Hewn Cabin   87

For the lower cut of the dovetail notch, align your template so it is even wit the upper cut and trace the triangle on the inside and outside of the log. Using your saw, ax or hammer and chisel, remove the triangle.

Plan where each electrical outlet will sit, inside and out, and all light switches and ceiling fixtures. Figure out the easiest and best concealed way to move the wiring from point to point around the room. Try to avoid running the wiring through corners as these can prove especially vexing.

You'll need to run your wiring through PVC conduit wherever it will run through the chinking. For my addition, I ran my wiring from the breaker box in my old house up through the ceiling and across the attic then down through the framed walls I put up to divide the space.

Where the space between the logs allows, try to run the electrical between the lower logs. When the wiring must climb the walls to reach a light or ceiling fan, try to hide it behind door or window trim.

If you are framing interior walls, take advantage of these. For instance, I ran some wiring through the attic, along the ceiling, then down through the framed wall to the gaps between the logs.

Whatever approach you take, work the wiring through the conduit as you proceed. Running the wiring through conduit in place and all sections attached can prove quite a challenge.

Plumbing is a little easier to address. If you carefully plan, there's no need to run pipes throughout the cabin. Try to concentrate all your plumbing in one area of the cabin, or in adjacent rooms.

Take advantage of the space beneath the cabin to get from point to point. If you live in a cold part of the country, make sure all pipes and connections are adequately insulated and protected against freezing.

A word of warning: Don't use wrap-around electric heat tape to prevent the pipes from freezing. It's a potential source of hidden fire, and in a log cabin that's the last thing you need.

The other key consideration about plumbing: make sure all your pipes and fixtures are accessible for repair and inspection. Hopefully you'll never need to make any repairs, but if you do you'll appreciate this advice.

As far as heating goes, You should know what type of heat source you will use and where it is best situated for even distribution of heat before you build. If you intend to rely on gas or wood heat, the most important consideration is how and where you will vent that source. If your vent pipes must pass through the walls or ceiling. Be sure to use insulated vent pipe to lessen the risk of fire.

Starting The Cabin

(Before you can start your cabin, you must decide how you will frame your floor. For information on the method I used, please refer to the Chapter 2 of Part III)

Once the logs are hewn and notched, building a cabin is a simple process, but it's also strenuous and dangerous work.

The cabin is put together by laying alternate pairs of logs, matching notched end to notched end for a secure fit.

Since the lowest logs bear the most weight, they should be among the largest and best supported logs in the cabin. If you're building on a continuous foundation, there's plenty of support. When building your cabin on stone or block piers you'll need to place these piers at both corners as well as in the middle of each of the bottom logs on each wall.

The piers or supports should be at least sixteen inches off the ground, and accessible for future inspection. They should also be slightly recessed from the edge of the log so they won't create a place where damaging moisture can gather. Further, as a precaution against termite infestation you can set a sheet of tin or other metal between the log and its point of contact with the pier.

The bottom surface of your sills which will support the entire wall need to be squared, as well as the sides for the interior and exterior walls. You will only need to notch the initial pair of sill logs on top, but the next set will need to be notched top and bottom. If the sills are to support your floor rafters you'll also need to hew the top surface of these logs.

With these four logs are place and supported, the first "course" of the cabin is done. Usually this is the stage in which you will frame the floor, or attach the sills to the floor deck which you built previously. If you did build the deck first, bolt the floor frame to the logs from its inside, using long lag bolts.

Framing The Door

Unless you're planning to cut the doors from the walls later, you'll need to build a frame for the door(s) in your cabin after the first or second course of logs is in place.

Whether you're going to build a plank door yourself or install a pre-hung door and frame, now is the time to create its opening. Take the dimensions of your planned door, or pre-hung door and frame, and build a box with inside dimensions at least half an inch larger in both height and width. Also remember to take into account the planned height of the final floor in the room, above the subfloor, when building the box.

Build this frame from two by six lumber, and nail a board diagonally from corner to corner on its outside to retain its shape. Before you can nail it in place, it's likely you'll need to either cut and hew an area to hold the frame on the log supporting it, or add a board or two to bring the door frame even with the floor. In either case, make sure the door frame is level front to back, side to side.

After you nail the bottom of the frame in place, use your level to get both sides of the frame plumb. Then, using scraps of boards of the appropriate length, make braces to keep the door plumb as the project proceeds. Nail one on the inside of each side of the door frame, and the other to the floor.

Raising The Logs

If you've got a tractor with a front loader at your disposal skip through most of the section. Otherwise you need to figure out a way to safely raise your logs progressively higher as you build the walls.

There's several ways to proceed. If you've got ample working space for a vehicle or a work animal, and a team of helpers you can pull your logs up a ramp and into place. That method was used to put up cabins in early communities where there was plenty of help.

It involves resting at least two logs or heavy beams on the wall to create a ramp. The builders would then roll or pull the logs up the ramp, and then set them in place. With several helpers this wouldn't be too much of a challenge, though it was potentially dangerous if the log or ramp slipped. A safer solution would be to hook the logs to a work animal or vehicle and then pull them up the ramp from the opposite side.

Since I didn't have the space, pulling power or extra hands, I had to figure out how to raise the logs myself. Drawing on my experience as a Boy Scout, I adapted the lashing methods I learned years ago to build giant tripods. I'll briefly discuss them here. (If you aren't familiar with lashing methods, ask a local scout or pick up a copy of the Scout handbook.)

Begin by cutting three poles of equal length for each tripod you intend to build. I used locust for mine, but any green tree at least four inches in diameter should be strong enough for the task. The poles I used ranged between 14 and 16 feet in length.

Set the trees on the ground alongside each other, using the largest ends for the bottom and narrow end for the top. Rest each of these poles on the same short piece of log or block about two feet from the top, so the ends you will be lashing are slightly raised and accessible. This will make it easier to weave your rope in and about as you lash the tripod together.

For the lashing you'll need a length of strong rope at least 15 to 20 feet long. Lashing begins by tying a clove hitch knot around one of the tripod legs, a knot which will not allow the rope to slip(see Scout handbook).

Once you've tied your clove hitch to a pole you begin lashing by wrapping your cord around all three trees, weaving in and out so it's tight enough to hold the trees together, yet loose enough to allow you to twist them apart to create the tripod. Finish the lashing, by wrapping your cord around the weaves between each tree, then tie the rope off with a clove hitch.

When the lashing is through, you can raise your tripod by gradually separating the two outermost legs slightly, and pulling on them so the tripod stands on its own. You'll probably need some help to get your tripods in the air and set up.

Spread the poles equally at the base until the point of your lashing is about 4 feet above where you want to raise your log. Once the tripod is in place, strengthen it by bracing its legs to each other and/or the base of your cabin. For braces, just nail boards from leg to leg, or to secure framework nearby.

Next, climb your ladder and secure a loop of rope or chain around the lashing where the three poles meet. Allow enough slack to attach your block and tackle. Lower the hook on your block and tackle and you're ready to raise the log.

Before you begin, always test the strength of the tripod and the holding power of the block and tackle. test the set-up by raising your logs a few feet into the air before you raise any logs over your head.

A single tripod like this will be adequate for raising sections of logs, but you'll need two, one near each corner of the building, for the full length logs you want to raise.

There's a hook on the end of the block and tackle for attaching to a rope or chain. I recommend using a length of chain around your logs to secure them to the block and tackle.

Once the chain is tied and secure to the block and tackle, it's merely a matter of making a series of short pulls on the rope to raise the log into the air. Always wear a safety helmet while doing this work and stand well clear of the tripods and logs. The block and tackle features a lock to prevent slipping, but if you must leave the rope unattended you should tie the end of the rope to some stationary object. That way, if anything slips the log will not come crashing down.

When you are working with two block and tackles , such as a full length log, it easier if you have a helper so you can raise both ends of the log at the same rate. If you are working alone, only raise the log a few feet at one end before tying the rope. Move to the other end and raise it slightly higher. Then continue the process, moving back and forth until you've raised the log high enough to set it in place.

The higher you raise the logs, the less working room there is. You may find that some logs drag or catch on the lower logs and notches as you bring them up. You can prevent this from happening by setting a stout board or section of log at an angle against the building so the log can slide up it as it is raised.

FRAMING THE WINDOWS

If you've carefully planned the cabin, somewhere around half way up you'll need to build boxes to hold your windows, and nail them in place. Measure the outside dimensions of your windows , where they will be set into the boxes. Using sections of 2x6 lumber build a box that is about half an inch larger, in its inside dimensions, than the length and width of the window. This will give you some "play" room when it's time to set your windows.

After the box frame is built nail a scrap of wood diagonally from one outside corner to the opposite corner, and cut it back so it is flush with the outside dimension of the box. This will ensure the box retains its shape once it is set into the wall.

The bottom of the box needs to be level, so it may be necessary to use a hammer and chisel, and possibly a saw, to make a level spot so you can nail the box to the log.

After you you've nailed the base in place, use your level to adjust the frame until it is plumb. Then, securely nail a pair of braces from both sides of the inside of each box to the floor. This will keep it stationary until you add the remaining sections of log around it.

Use several large nails, at least 16 penny, to nail through the frames into the of ends of the log sections surrounding it to hold them in place.

THE CAP LOGS

Raising the rest of your log walls is pretty straightforward, and involves no additional work until you get to the top or "cap" logs.

Each pair of these logs, on opposite sides of the building, should be roughly the same height. Before you can nail your ceiling and roof rafters in place, though, you have to work on the logs a little more to make sure they line up exactly. If your walls are an uneven height, it will slow the process down, as you'll have to cut each rafter individually. So it's really important that you get these logs even with each other. This will probably mean some more work with a saw, chisel and plane.

The rafters for the ceiling and roof usually run parallel to each other. That way, you can nail them together at the plate and to the wall, and strengthen the structure.

Decide how your roof will run, where it will peak, it's slope or pitch, and how the roof ridgeboard will run. Front to back, or side to side? All roof rafters will be nailed perpendicular to the ridgeboard board and parallel walls.

When you know where the rafters need to be nailed, compare the height of the logs on each side of the building. Attach a line near one end of one of these logs at its top, and stretch it across the room to the same point on the opposite log.

Attach a line level to the line, and move your line up and down until the bubble settles in the middle. Play with the line and line level, moving your line up and down so its as near to the top of both logs as you can get it. You want to adjust your line so there's the least amount of work involved in adding to or hewing down the logs so they are even with each other. Once you've determined that point mark where the line touches both logs.

Repeat this process at the other end of the logs.

You may find you have to readjust your line, up and down until you get a range which is workable. In extreme cases, you may need to build one of the logs up, by nailing a 2x6 to it to serve as a plate, while shaving the top of the opposite log. When you need to remove part of the log you can adapt the scoring method used when hewing the logs.

If possible, use your square or level to mark a straight line from the inside to outside at the end of each line in the narrow gap between the notch and top log. If the notch is too tight to permit this you'll have to approximate, then make adjustments later. Once you have those marks made, stretch and snap your chalkline from mark to mark on the inside and outside of the log.

Using these chalklines as your guide, make a series of "score" marks. Cut down into the log to the depth of these lines with your saw. Use a hammer and chisel to remove the top of the log as needed.

Working with a chisel, and possibly a plane to smooth the top of the log off, this work can take several hours before both logs are completely level and even with each other. You'll appreciate the time spent when you start framing the roof.

(Before you can frame the roof, you need to frame any interior walls and ceilings if you are using standard construction. For an outline of what's involved consult the appendix on Adapting the New at the back of the book)

THE ROOF

Please accept this practical advice. If you've never built a roof before, do as I did and hire someone to help with the framing. You can put down the lathe and tin or shingles yourself, but framing the rafters, making the right measurements and cuts, requires some expertise. This is not the place to experiment. If the roof isn't done right, and it leaks your cabin won't be a happy place on rainy days.

In fact, the roof may be the most important part of the your structure. It's protection impacts both how much you can enjoy living in the space, and its longevity. You've got to get it right, pure and simple.

If that means hiring someone who knows how it's done, by all means do.

If you plan to do it yourself, consult a book on framing or basic carpentry for detailed instructions on how to determine the pitch and span, use the framing square, and make the right cuts for the ridgeboard and the plate.

Here's a basic outline of the steps involved in planning a typical gable roof, the type most often seen on cabins. (Incidentally, original log cabins were built with very steep slopes so rain would roll off the roof before it had the opportunity to slip between the shake shingles. Unless you are building an authentic shake roof a more gradual pitch will serve fine.)

Planning the roof begins with determining the span, or distance between the outer walls of the structure where the roof rafters will rest. In the standard roof you would divide this distance in half to determine the location of the ridgeboard. How high that ridgeboard sits above the structure depends on the desired pitch of the roof, or distance the roof rises for every foot of run. The height of the ridge board will also determine the length and cut of each rafter. If you're planning a porch as an extension of the roof, or an overhang, these must all be taken into account when figuring the total length required for your rafters.

Before you can attach your rafters, you have to position and support your ridgeboard in Place. Use a pair of 2x4s or 2x6s, nailed to the board and each end of the cabin.

Once you have the ridge board in place, and know your pitch, there's two key cuts to be made: first, the "plumb cut", where the rafter meets the ridge board, and a second, called a seat or heel cut, where the rafter meets the plate at the top of the wall.

You can determine the angle of these cuts, as well as the length of your rafters, using a framing square and the tables etched along its surface, if you know how. Again, I strongly recommend that you consult a book on carpentry. I don't want to give advice which could result in your leaky roof. There you'll find information on how to use the framing square to determine your cuts, as well advice on variations on the simple gable roof.

Experienced carpenters don't even need a square. They can hold one rafter up, make their marks and cuts, and then use it as a pattern for the rest of the roof. This takes practice. And this is where you'll appreciate all the work that went getting your cap logs flat and level with each other: if they aren't you can't use a pattern and will need to mark and cut each rafter individually.

If you are using a pattern, check your first rafter, after making the cuts, all along the ridgeboard and plate. It should be a good fit everywhere. After you've determined it is, mark and use it as a pattern for cuts to be made on all other rafters on that side of the roof. You should make a separate pattern for the rafters on the other side of the roof.

Once the rafters are nailed to the ridgeboard and top logs, you need to put down the lathe or sheathing to which you will attach your felt, if used, and roofing. While 4x8 sheets of plywood are used as sheathing in most modern homes, many log cabins use evenly spaced boards of 1x6, 1x8 or 1x 12 lumber for lathe under a tin roof.

I can only recommend a tin roof because that's what I've always chosen for my log buildings. Tin goes up very quickly, and done right it has a long life. For an amateur like myself it's a quick and workable solution.

And, there's something to be said for the sound of rain falling on a tine roof. It can get loud but I find it a soothing, almost musical source of white noise.

If you've never heard it, by all means please do before you decide how to roof your cabin.

Final Steps

With the roof overhead, you can start focusing on all those chores which make the cabin a home. Depending on how ambitious you are, and how fancy you want the cabin to be, these can take as much time as building the cabin itself. For this final phase of the project you can adapt and employ many of the modern building methods and materials briefly discussed in Appendix B

Once the building is "in the dry" you can finish the work on the electrical and plumbing, pulling the wires, setting the outlet boxes, switches and lights, installing pipes.

If you want to treat the logs to bring out their grain or protect them against the weather's, it's advisable you take care of that, too, before you begin work putting in the chinking.

Also, insulate the floor and ceiling as needed.

Complete any work required on your heating system, such as the construction of a hearth and chimney, installing vent pipes for your heat source. When your house is made of wood you want to err on the side of caution and fire protection.

Applying the chinking can be slow work — after filling in the gaps between logs with small stones to hold the "mud."

For the chinking, follow the same procedures outlined in Chapter 1, Part II. Be aware, though, that your logs will continue to shrink for several years after they have been put up. Check on the chinking at least twice a year, and touch it up as needed with more cement or clear silicone caulk.

After the chinking is done, and you've cleaned up the mess, finish the walls and ceilings as needed with drywall, and paint them.

When all the other work is done, put down your floor over the subfloor, and finish or cover as desired.

Install interior doors and trim, and attend to whatever else needs to be done.

I'll bet you're already moved in well before you've finished with the final touches…if you're ever finished.

PART III

Building A Cabin With Round Logs

1

A "Quick and Easy" Cabin

After completing the hand hewn log addition to our house I swore off building any more log cabins…for a while.

I'd been working out of a corner "office" in our large upstairs bedroom as a freelance writer the entire time we were building the log addition. It was tough on me, but tougher on Liz and our four kids to always have me around, begging for some measure of silence so I could focus on my work. We managed to make the best of a difficult situation.

It might have worked a little longer, but in early 1993 we found out, much to our surprise Liz was expecting again. With our fifth child due in August, we needed my office space for the baby. I had no choice but find other working arrangements.

Money was tight, and with the expense of another child looming, I started to explore the possibility of raising another log cabin. It seemed the most affordable approach to building and there remained enough tall straight poplar trees scattered around our place to give me enough logs.

My concerns soon shifted from "if" to "how" I should approach the project. I knew one thing: I did not want to put myself through the chore of hewing out the logs. I wanted something I could put up over time relatively quickly, working as I could, without the intensive labor involved in hewing the logs.

I finally decided the quickest and easiest solution would be to build this cabin from poplar logs as round as the day they were cut, but stripped of their bark. To hold the corners together, I planned to use rounded saddle notches, cut with a chainsaw.

As for the location, I wanted something close enough to be a convenient walk from our house, yet far enough removed that I could physically and mentally escape work at the end of each day. After scouting possible sites, I decided on a small semi-flat area at the top of the ridge opposite our house and above our barn. It's the spot where Liz and I spent our first night on our property huddled in a tent. There's the remnant of a wagon road which climbs about 300 feet up a steep hill past this spot to an old slave cemetery nearby. I figured the hike up that hill, several times each day, would be good exercise for my heart and legs.

I already knew, first hand, cabin projects tend to take longer than anticipated, so I also looked for a short term solution for my office needs. I ended up spending $100 for a much used camping trailer which had been left behind on the farm of some friends. Carlos even hauled the trailer up the hill with his four wheel drive truck and helped set it up in a small clearing I cut in the woods. It was within 100 feet of where I planned to build my office cabin, so I was able to power the trailer from the temporary electric service I had set up for the project. I ended up working in that 8x12 trailer for the better part of two years while toiling on my "quick" log cabin project.

While it served as my office I only experienced a few problems. Even in the shade, it was tough to cool the trailer in the summer and the thing was almost impossible to heat in winter. On especially cold days there was little difference between the temperature inside and outside, once you moved a few feet from the electric heater kept under my desk. It got so cold at times, the tape on my answering machine barely turned!

I never could figure out or find the cause of the other problem, which concerned the electricity. In wet weather, I'd get a good shocking jolt from the brass door knob whenever I tried opening it from outside. I figured the short was caused by a frayed wire touching the metal shell of trailer, or an improper ground, but I tore apart the trailer walls in search of the cause without success.

Eventually I came up with a low-tech solution: I'd just wrap my shirt around my hand for insulation when opening that door on damp or rainy days to avoid the shock.

It wasn't the best arrangement, but it served as a "temporary" office. I even wrote my first book in those cramped quarters. And all the while, I only had to look out my desk window at my cabin in progress, and dream of the days when I would actually be able to work in larger, more comfortable quarters.

I had decided a square cabin, a 16x16 room, would give me enough space to work, relax and receive the occasional guest.

The toughest part of this project proved getting the logs to the site. Yes, there were plenty of usable trees, but I'd already used all which were easily accessible for the log addition to our house. What remained were all standing down steep hillsides, some growing as much as 2oo feet from the top of the ridge.

Here's where having a friend with a 4-wheel drive really helped. I cut down the trees in late fall, and had rolls of galvanized cable discarded by the phone company to use as tow lines. Several afternoons after we cut firewood that winter, a group of us—Carlos, his son, a few of mine and myself—backed his 4-wheel drive truck to the edge of the ridge, strung together lengths of cable, attached them to his truck and dragged the logs up the hillside. We moved about half the needed logs to the building site before the truck broke down, permanently.

An alternate solution soon presented itself when a friendly old farmer, Mr. Bob, stopped by our house and asked if I was interested in his old truck. It was a red 1963 Ford F-250 with a "granny gear" transmission that could pull the hills. After hearing of my plight, he'd gotten it running after years of setting in his yard. Since it smoked like the morning fog he only asked $200. I figured as long as I could keep it in oil it would get the job done.

I could also justify the expense as I now needed something to haul our firewood. And the truck, rough as it was, still held some nostalgic appeal. The very first vehicle Liz and I owned was a 1963 red Ford F-100, almost its twin except ours was a "unibody" (the truck bed was attached to the cab). So I bought Mr. Bob's truck intending to use it and one day restore it. I only need look out my cabin window today for a constant reminder of that plan, as the truck sits, its work done, partially dismantled in a special area in our woods devoted to planned "classic car" projects.

Anyway, that old truck chugged, smoked and crawled along the ridge in low gear, snaking the remaining logs up the hillsides and through the woods to where I stacked them in preparation for building that cabin.

All the while I was gathering my logs I was also looking around for salvageable doors and windows for my planned office. With money spent on the truck what little budget I could wrestle had to cover the cost of lumber for the floors, ceiling drywall, tin roofing, cement for chinking, and everything electrical. Log cabins tend be dark inside but I like a lot of light so I hoped to round up two windows for each wall as well as a front and back door. (Looking back, I don't know why I worried about a back door in such a small cabin, but it served me well later.)

I had some windows left over from our old house, and neighbors donated an odd sampling of other windows they had laying around their sheds and barns. All I had to do was rebuild some of the frames, replace a few glass panes and they would work fine. For the front door, I reclaimed the old back door from our house, and saved another which was on its way to the country dump from our church, for use as a back door.

Since I already had an "office" I wasn't under the same pressure to complete this project as with the addition to my house. That was refreshing, and so was the solitude I enjoyed while working on this cabin. Well removed from the road, the site didn't invite the parade of curious which went with working on our house.

A few friends hiked the hill for a look-see, or to help for a day, and our eldest son Marcus pitched in throughout the project. The entire family, Liz, the four older kids and new baby Matt, came up for a picnic lunch on the hill a couple of times. But most of the time I worked alone, taking my time, thoroughly enjoying the process and progress.

Once again my plans began with building a freestanding deck, this time 16x16 feet. This would serve as the floor, and a frame I could raise the log cabin around. I used rough-sawn poplar lumber for framing this entire project, cut fresh and green by a local saw mill. That green lumber worked fine as long as I used it quickly or kept it properly stacked. When I didn't, and left it to dry in the sun I had to work with a severely twisted or warped timbers.

I began by building piers, eight feet apart, high enough that my sills would sit at least 16 inches off the ground. On top of the sills, made of 6x6 poplar beams, I framed the floor using 2x6 polar rafters. For the subfloor I again used cull lumber of varied types, widths and lengths. With the floor deck complete, I focused on starting the log walls.

The rounded log method I'd chosen for this project greatly simplified building of the cabin, and the number of tools required. Building with rounded logs, and setting them in place using a rounded saddle notch, is about as simple as building a log cabin can get. It did take some time to develop the skill required to control the chainsaw to cut smooth notches. But even on the first few logs, where I did a rough job of it, I managed to get a close enough match that holds fine.

The system I devised—and I'll grant that there are probably better methods involving less work—required first that I set each log in place on top of the log I would be notching it to. Then I could mark it for cutting. With the rounded saddle notch, only the top log is notched on its underside. The notch fits over the log below it, like a cup, so any rain or moisture will naturally roll off the log.

Although it was supposed to be a quick project, the rounded log cabin took shape slowly, over several seasons.

My method began with marking the log for its notch cut. Using a standard compass loaded with pencil, I set it to the desired depth of the notch, usually about three inches. Then, I traced the shape of the lower log with the compass, while the pencil inscribed the shape on the log to be notched.

After marking the log this way on both sides of the notch, it was simply a matter of setting the log on the ground, and flipping it so the side to be cut faced up. Then I'd anchor it in place and start cutting with my chain saw, using these pencil lines as guides. First, I'd make a series of close cuts, two to four inches apart, to the lines, and then knock the wood loose with a hammer and chisel.

Then finish the notch with the saw. I was working with a large saw with an 18-inch bar, and most logs were about 12 inches in diameter or slightly larger. I messed up a few times in the beginning as I tried shaping the notch. After a while, though, I learned how to control the saw to shave just what I needed.

I believe this part of the project would have moved quicker, and the results neater, had I been working with a lighter, more manageable saw. I think I might have spoiled a log or two in the process because the large saw cut so quickly. Other times, it took several tries at trimming, and matching the notch to the log below, before I had an acceptable fit. Once I did, I set the log in place spiked the corner with a 60 penny nail.

As soon as the first four courses of logs were in place, with gaps for doors in both the front and back walls, I worked on frames for the windows, scrapping them together from odd pieces of 2x6s and cull lumber left over from the floor. Before I could nail them in place, I had to chisel out level areas where the boxes would sit on the top of each of the round logs. The idea was to nail the boxes in place, then nail braces on their sides tacked to the floor to keep them plumb.

I apparently wasn't thinking the day I set the largest of these boxes in place, for the double front window, without nailing it. Soon as I turned from the wall I was alerted to my mistake, and fact that I wasn't wearing a helmet, when this heavy box came crashing down, around me, brushing my hair as it fell. You can only make so many mistakes like that…my guardian angel must have been watching, as my mother used to say.

I worked as I could on the cabin through the fall. The cold weather was soon upon us so I took a winter break from the project during January and February. Before I did, though, I made sure all the unused logs were stacked well up off the ground where they couldn't be damaged by moisture or dampness.

Every day I passed the cabin on my way to work I was reminded of the work that awaited me with the arrival of spring. Come March, with the woods awakening and the days growing longer, I shook the dust off my tools and resumed the project.

Now that the window boxes were in place, I could start working with smaller sections of log. Still, some pieces were as long as six foot and weighed more than I wanted to try to manage by lifting them into place. Again I resorted to my proven tripod method for raising the rest of the walls. I cut 6 locust poles about 14 feet long each and lashed them together in sets of three for use as tripods.

I used the tripods and my pair of block and tackle on the rest of the logs, including the full length 18-foot logs which capped three of the walls. By the end of May the walls were up and almost ready for the ceiling and roof rafters. It took a couple of hours work with the chisel, saw and plane to flatten out the front and rear logs, level with each other, so they could serve as the plates for nailing down the roof rafters.

When I planned my office cabin, I decided I'd seen enough roofs go up that I could figure out how to do it myself. It made sense at the time, but I realize now I may have been a little too confident in my unproven skills.

My plans included a dormer at the front of the building , with two windows, as a way of inviting even more light into my office. My friend Drew donated a day's effort to help me frame the roof and dormer, and by sundown the roof rafters were all in place. Marcus helped me with the lathe. Later in the summer, when I had money for the tin, it took only an afternoon for us to put down the roof.

I realized that day something wasn't right with the dormer when we had to make so many cuts to make the tin fit so the dormer roof would shed water. Looking back, I realize I should have ripped the whole dormer out and redone it, or had someone who knew what they were doing reframe it for me. I didn't though, and have more than paid for my stubbornness emptying pans from the intermittent leaks which reintroduce themselves at its corners from time to time. Live and learn...

Regardless of the leaks, with the roof over head, I set the windows and doors in their frames. Now it felt like a cabin. Now came the job of wiring the place. I'd learned my lessons at our house, and had carefully planned ahead how to run my wires this time round. I divided the room into quarters, and ran the wiring through conduit under the cabin, up into each corner and then along the wall in the gaps between the logs.

Each wall is served by its own breaker, and I used additional breakers and conduit for the porch and ceiling lights, and the power supply for my desk and computer equipment. It took about a week's worth of weekends to run the wire, before I was ready for the initial "rough in" inspection.

It did pass, and chinking the gaps between he logs became my priority. For mud, I used the same combination of Portland and mortar which had served me at the house. My portable cement mixer again saved significant time and labor, but offered no relief from the tedium of working the mud into the spaces separating the logs. It's still the part of cabin building I least enjoy.

By now it was late October, and another cool season was approaching. On days when the potential for a freeze could ruin an afternoon's worth of chinking, I concentrated on insulating the cabin and putting up a drywall ceiling. I took what warm spells there were through that mild winter to finish up the cabin: put up trim, insulate the roof and ceiling, finish and paint the drywall and wire the individual outlets, switches and lamps, and put down and paint the floor.

In late March, with all the work done, I passed the final inspection on the electrical.

The first week of April I moved out of the trailer and into my new office.

It's a very nice space.

My desk, computer and office equipment claim one full corner of the back wall. The chair where I am now seated faces south and west. I can look through the double window on the front wall at an unspoiled panorama of hills, and hollows, stretching for miles. To my left or right other windows offer a glimpse of the near woods.

The local gas company actually ran a line up the hill to my office so I have heat in winter at the turn of a dial. Since there's electricity, I can enjoy an air conditioner on hot summer days, and any other amenity of modern living. My guitar, mandolin and banjo wait in a corner for when I feel like strumming a few chords. There's a carpet on the floor and a couch, chair and rocker for receiving guests. I figure I owe at least a few minutes of my time to anyone who makes the trek up the hill to see me here.

A 16x16 cabin makes a generous office, and something of this size, or a little larger, would make a comfortable retreat for the individual, couple or small family. It's simple to build, and can go up quickly if you've got some help, and the materials on site.

The first time I sat in my desk chair I looked around my office, and thought, quite content, that it marked the culmination of my log cabin projects, but...well, you never know.

The completed cabin, built of rounded logs, serves as my office.

2

Methods for A Rounded Log Cabin

When the pioneers needed a first home they cut the trees, trimmed the branches and started piecing together a cabin, from the ground up. It was a quick solution for those with an immediate need for shelter.

I decided on building a rounded log cabin for my office as a way to save time and eliminate the work involved in hewing and notching the logs. I did save a significant amount of work, but it still took a while before the cabin was ready to move in. Almost two years in fact. I only worked on this project on and off, as I could. Most of my time during this period was taken up with eking out a living as a freelance writer and the many chores that go with living on a "farm."

If you have the help, tools, machinery and logs, you could probably put up the shell of a comparable rounded log cabin in a few days. But it would still be a few weeks before you have the wiring, roofing, chinking and finish work out of the way.

You will escape the labor involved in hewing the logs, but can't avoid the many steps involved in building any type of cabin. Much of this is already covered in Part II, Chapter 2 so I'll briefly summarize the procedures here. The rest of this section focuses on what's unique about building with rounded logs.

A rounded log cabin has a distinctive style and look which most people think of first when they imagine a log cabin. Since you don't need to cut away the outer lay of wood to create a flat face for the cabin walls, you can use much smaller logs than in a hand-hewn cabin. In areas of the country where pines are readily available cabins can be built using trees as small as six to eight inches in diameter.

For my cabin, built of poplar, I used logs ranging in size from eight inches to a foot in diameter. The inside ceiling is about 7–1/2 feet high, and each wall is about 10 logs high.

Planning

Begin, as with any other cabin by planning the size, focusing on the inside dimensions of the room Figure out the size and locations of windows and doors, and the boxes you will need to build to hold them.

Draw up a diagram of the cabin, including the location of doors and windows, and the height of each wall. Based on these plans you can estimate the number of logs you'll need for each wall and the entire cabin. Decide on what diameter logs you'll use, and estimated gap you'll leave between each log, if any.

With this information, and your diagram, you should be able to estimate how many logs are required for the project. You'll want to cut at least 10 percent more trees than the total number needed to make sure you have enough logs on hand for the total project.

During initial planning, it's also prudent to decide how and where you want to run your electricity and plumbing. Deciding ahead on the placement of pipes, wiring and conduit will save much work and frustration if you leave these considerations until later.

The same goes for your intended heat source. The need for and placement of a hearth, chimney or flue, will impact your plans and the layout of the cabin. Better to think ahead than try and improvise later.

Building any cabin with frames in place for windows and doors speeds the work along and reduces the number of full length logs required.

CUTTING YOUR LOGS

You'll find it easier to work with logs after they have had a chance to dry, but a rounded log cabin can certainly be put up "green." If you want to work with lighter logs, though, ring the bark during the growing season as they stand or cut them in the late fall/early winter, when they hold the least amount of water. Either way will help reduce the amount of moisture in the logs, and their weight, before you start working on them.

Once the trees are down, cut your logs from the trunk, trim them of branches, and prop them up so they are not in contact with the ground while they dry. Check on them, and rotate them side to side as needed, until you are ready to build.

Prepping The Site

Since I did not want to try and build the floor into the log walls, I began each of my cabins as self-standing deck, the full inside dimensions of the room. I then raised my cabin around the floor, attaching it to the lowest or sill log with long lag bolts.

There are several steps involved:

Build a series of piers or pilings from rock, block or poured concrete foundation to support the sills every eight feet. These supports should be at least 16 inches off the ground to minimize the potential for infestation by termites.

For my 16x16 foot cabin I built one pier in each corner, and another midway between them along the outside wall. All supports along the outer wall were large enough to support both the floor sill, and the sill or lowest log of the cabin wall. I used at least three courses of two 8x8x16 cement blocks, filled with concrete. With another pier in the middle of the cabin, I set three 6x6x16-foot sills, 1 every eight feet apart, to support the floor rafters.

For framing the floor I used 2x6x16-inch poplar rafters set on 16 inch centers, perpendicular to the sills. Once the floor rafters are in place and nailed to the sills you can lay down a subfloor the full dimension of the room. I used one inch cull lumber of varied lengths and widths, but plywood, at least 1/2-inch thick will also work. Avoid OSB or wafer board for the subfloor as any it will weaken over time if exposed to moisture.

With the floor platform built, I was ready to start with the logs.

Please consult Chapter 2 of Part II for considerations on moving logs to your site, raising the logs, the construction of tripods, building frames for windows and doors, and considerations about the roof.

THE ROUNDED SADDLE NOTCH

Once you've stripped the logs of their bark, the only real work to do to them is cutting the notches. For my cabin I used a rounded saddle notch, and any variations of the saddle notch would also work fine.

If all your logs are the same size you can work from a template. Simply make a pattern representing the size and shape of your notch, trace it on both sides of the end to be notched, and cut away!

If you're using logs of varied sizes, the toughest part about cutting the rounded saddle notch is that it requires raising the log and setting it in place in order to mark it for notching.

With this type of notch I only cut the bottom of the log, the part that would fit over the log beneath it. When marking the trees for the notch you want to be careful not to cut the notch too deep or the log won't be usable. Almost all the notches used in my rounded cabin ranged between three and 3–1/2 inches deep.

Once you have your log raised and resting in place, marking it for the notch is pretty simple. Take a standard school compass loaded with a pencil, and set the compass for the depth of the notch you want as if setting it to trace a circle.

Start on the outside of the lower log, holding the compass steadily so the pencil is at the bottom of the upper log, and directly above the compass point. Move the compass point slowly up over the top of the lower log so the pencil inscribes a half circle across the lower surface of upper log. This arc should pretty closely resemble the shape of the top of the lower log.

Repeat this process on the inside of the log.

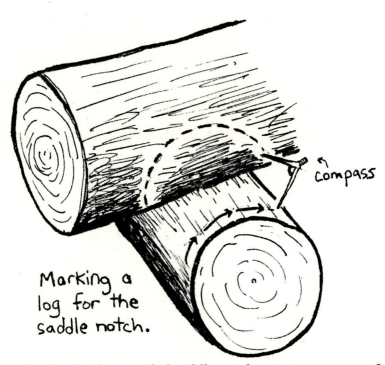

Marking a log for the saddle notch.

To mark a log for the rounded saddle notch, set your compass for the desired depth of the notch. Then, trace the point of compass along the upper surface of lower log, allowing pencil to enscribe this shape along the log to be notched.

Done right, both sides of the log to be notched will feature a well defined semicircle marking the area which needs to be removed.

Next, flip the log over and wedge it in place so the circle is on top of the log. When working on the upper sections of the wall it may be safer to lower the log to the ground, where you can work it without worry about it falling.

Depending on the size of the notch, and your own skills, there are several ways to cut it. If you're an experienced user of a reciprocating saw with a long enough blade you can cut the area out. A small chain saw can be used for cutting the notch, if you can carefully control the cut. It does a rougher job, though, and you will need to do some touching up.

I improvised as needed, resorting to some combination of chainsaw, hammer and chisel and small hatchet to get the right fit.

If cutting the notch by hand, use the lines drawn with the compass as a guide, and begin by cutting several score marks defining the area of the notch. Then with the ax, or a hammer and chisel, cut away the wood to make the notch. At times, I used my chain saw, shaving back and forth to smooth the cut but I wouldn't recommend this. The chisel is slower but safer and works fine.

Once the notch is cut, set it in place. Since no two logs were exactly alike, and each notch therefore different, I had to refine some of these notches as many as three times before I had the right fit. You may need to use your pencil and compass to mark the log a little deeper, and then repeat the process of shaping the notch for a tight fit, to reduce the gap between logs.

When it's is right, the log should fit neatly over the log beneath it, gripping it with a curved cup. For added strength I spiked every log to the one below it using 4o or 60 penny nails.

THE REST OF THE CABIN

Other than notching, raising the rounded cabin was really no different than working on any other type of cabin. If anything, it's much easier. You need make sure your window frames and doors are braced so they are both plumb and level as your wall goes up.

When you get to the cap logs, You'll still need to take the time to level logs on the opposite walls to serve as the plate for your roof rafters.

From there it's like building any other building, as discussed in Chapter 2 of Part II and the "Adapting The New" Appendix with one exception:

122   A Home From the Woods

For the rounded saddle notch, mark the upper log with the shape of the lower log, and remove a cup shaped section from its bottom. Make the initial cut with your saw, clean it up with a chisel, set it in place and spike it for a secure hold.

The Chinking

You can use the same methods and ingredients for chinking as outlined in Chapter 2 of Part I.

With rounded logs, though you have to take some special precautions.

On a hand-hewn cabin the rain tends to run straight down the face of the logs, and drip onto the log below it. With a rounded log, the rain rolls with the contours of the log. If the chinking doesn't tightly seal the log in place, the rain will gather behind it, making the log susceptible to decay from moisture collected there.

Rounded logs, since they are full sized, can hold more moisture than a hewn log, even when given the opportunity to dry before you start building with them. Once the logs are up and standing, they will continue to dry and shrink. This will loosen the chinking along the edges, as well as introduce deep "checks" over the surface of the log.

These slits, can run along the face of the log for several feet and penetrate several inches into it. They provide a natural place for rain to gather inside the log, or for insects to start building a home.

To prevent these potential problem area, you must continually monitor the conditions of the logs for several years after putting them up. When you discover a potential problem area, patch it with more chinking or silicone caulk.

I personally recommend the silicone as it's easy to work with, easy to clean up and remains somewhat pliable for years.

Part IV

*Removing and Rebuilding
An Original Log Cabin*

1

Trials and Tribulations With An Unexpected Find

The last thing I expected as I headed out to empty the trash one fine August day was to come home with a log cabin.

But that's what happened…sort of.

We'd already restored one and built two log cabins. I couldn't imagine what else there was to learn about them. As a consequence of my interest and experience with them, I had investigated all the surviving cabins in the area I could find.

So my curiosity piqued when, as I drove down the road, I spied the old white farm house and saw the first signs that it was about to be dismantled. Two people could be seen breaking down its tall chimney, and a few of the white clapboard boards had been pulled loose around the front of the house.

I knew the place, and its history well. I'd always heard it was haunted. From the outside, it presented itself as the typical two story white clapboard farm house found in this area. Within those walls, though was a log cabin, larger than my own, and one with some local lore. The thickness of the walls in the main room offered a hint of the cabin within, but the only place logs were visible was in the attic loft, and the cramped staircase which led to it.

That cabin had been the ancestral home of one of the first families to settle this area. The last family member to own it had done his best to keep the place up as a monument to the family roots throughout his life. When a new highway was built near its site(a hand dug well still identifies the original location), he'd had the house moved across the road. I even have a Civil War map of the area identifying the location of the house.

The old man was proud of the homestead. Late in his life, a newspaper ran a feature in which he talked about the place, its history, and all it meant to him. When he died he left it to a relative, apparently hoping the house and its heritage would mean as much to him as it had to the old man.

Well, within a year or so the house and homestead were sold. The buyer seemed to care little about the place or its history. In fact, the local scuttlebutt held his real interest in the place was as the potential site for a future subdivision. Anyway, after a couple of years under his stewardship a portion of the tin roof was blown off one end of the house in a storm, and never replaced.

I'd shake my head each time I passed, knowing part of that cabin was doomed to the elements. It stood, hidden within the larger house, as one of the few surviving original cabins still intact in our area.

When I saw it about to be dismantled I assumed it was being taken down for the logs. Over the years here I've seen several cabins taken apart so the logs could be used in new buildings. We'd even had someone knock on our door to inquire about purchasing our logs after Liz and I had restored our cabin, and were living in it!

So, I parked my truck and walked over to the old house. It was our neighbors Lori and Larry who were dismantling the chimney for use in their camp. They directed me inside to the men who were responsible for the house.

I found both at work with crowbars, loosening boards from the wall in a room adjacent to the portion holding the cabin. They'd already made a haphazard stab at removing some of the boards covering it, enough to reveal scattered sections of logs.

"Taking down the old place?" I asked. They nodded something to that effect, somewhat uneasy at this stranger's arrival.

"So what are you gonna do with the logs?" I asked, being nosy.

They didn't answer right away, then one, the shorter and thinner of the two, set his crowbar down. He spoke for both.

"Well...what'll you give for them?"

His response was totally unexpected and caught me off guard. My thoughts raced: did I need this?...what would I do with this?...what can I really afford?

"I don't know," I paused. "A couple of hundred dollars."

He turned to look at the other man, who nodded. "Buddy, it's yours," he said.

"Two hundred dollars," I repeated, quite shocked.

"Two hundred dollars, down and stacked...in your yard, if you want."

Quickly I extended my hand to shake with both of them, repeating "Two hundred dollars." That's all it takes to close the deal around here.

They were quite happy with their sale, and I couldn't believe my luck. I rationalized to myself that I could tack the cabin on to the back of my office as another room. I thought it would be another "quick and easy" solution, and give me a chance to take down and rebuild an original cabin.

That's something I hadn't thought lacking from my log cabin expertise until then.

I offered to help them dismantle the entire house, and agreed to start the following afternoon.

When I got home I honked the horn of the truck until Liz and the kids came out on the porch.

"You know that old log cabin, the one inside the white house?" I smiled.

"Well, I just bought it."

We had agreed to meet the next afternoon, around 4, but they were late. They seemed a bit uneasy to find me already there, perhaps worried I might have taken something of value. They introduced themselves as Carl, the older spokesman for the two, and Gerald his brother, a bit taller, bald and round. They both seemed to me to be in their mid to late 40s, while I was just pushing 40.

"The man what owns this place said we could have it all, as long as we tear it down and clean it up as quick as possible," Carl explained.

Gerald added they had intended to take down the logs and use them for a retaining wall to hold back a bank of dirt in the driveway of their property. "What's left, I's just gonna cut up for firewood," he added.

I told them about myself, where I lived, my experience with log cabins, and what I knew of the place. "I always heard it was haunted too," agreed Gerald, with a smile and nod.

I asked that I be allowed to keep everything from the cabin, as I was planning to rebuild it as close to the original as possible as an addition to my office(which wasn't yet near completion). While waiting for them that afternoon I had found a pair of hand-hewn frames they had pried loose from a front window. They had split one, but both sections still held long hand-carved hickory used to anchor the frames to the log wall.

"Imagine having to make all your own pegs," Carl marveled.

They agreed to let me have everything to do with the cabin except for the mill cut 2x8 beams visible through a gap in the ceiling.

Since there was much work to be done before the cabin could be taken down, I again offered to help them with their work. I felt they had given me a great deal, and it was the least I could do. It also gave me the opportunity to be there and make sure nothing happened to the cabin.

They wanted assurance I didn't expect anything more than the logs, and were then happy to accept my offer. For the next couple of hours I worked alongside them, removing the boards which made up the walls in the rooms adjoining the cabin.

Around six they explained they had to leave for supper and would be back tomorrow. I told them I was going to hang around a little longer to take some pictures and measurements of the cabin. They agreed and left, but their slight suspicion showed.

The old house, and the house within.

First thing I did was take out my camera and photograph the outside of the house from all sides. It still looked like the old white farm house, mostly, but where the weatherboard had been removed, you could glimpse some of the interior and sections of logs.

Back inside as I walked from room to room around the house I could discern some history of the structure. The one room cabin with its chimney at the far end had served as the focus of family life since it first went up. I tried to imagine it, alone and isolated at the edge of the wilderness in the early 1800s, when settlers were first pushing into this area.

That one room had three doorways and four windows: two on either side of the door, and two more on either side of the chimney. There was a doorway on the back wall and another on the wall opposite the chimney. That door led into a hallway, with another door. It opened to a small enclosed staircase climbing the outside the wall of the room.

The stairs led to a loft, three logs high, which probably served as sleeping quarters. I later learned the logs used in the loft were of a slightly different style than the rest of the cabin. That loft was probably built in the initial effort to add more space to the one-room cabin.

I figured the front door and window frames were original, as all three were built of hewn lumber held in place with carved hickory pegs. I also figured the other doors were originally windows which had been enlarged into doors as rooms were added around the cabin. Later, when the house grew again, these were converted into interior doors, the two windows around the chimney were cut into the walls to let more light into the room.

Walking through the house, I decided the first downstairs room added was a little kitchen area, with its own chimney, just off the door opposite the front wall. This, too, was later expanded, when sometime in the late 1800s an entire new wing of the house was added. It completely surrounded the original cabin and stood two stories tall.

This new house was covered with white weatherboard siding, and had a new roof and porch. This large white home was what the old man had known his entire life. Its cabin roots were entirely hidden, except for the few exposed logs in the staircase wall and the upstairs loft. For generations the cabin room had served the family as their parlor.

Most of the cabin remained hidden beneath drywall, which had been tacked to layers of paper used to hide the humble origins of the cabin long ago. I took many pictures that afternoon, but they reveal little of the cabin within that house.

Then I put down my camera, and picked up my tape measure and stepped into the old room. There was an unexpected chill, and it reminded me of the stories I'd heard about cabin ghosts. I reasoned, though, it was more likely the cabin was so well insulated that it always stayed a little cooler

Beneath the drywall the walls held thick layers of old wallpaper in many patterns, and under them old newspapers and catalog pages. In the rare patches where the paper had peeled away from the logs I could see they wore the thick whitewash coat which had covered our cabin at home.

The room measured 15x23 feet inside. From what I could see the logs ranged between six inches and eight inches thick. When built, this cabin measured 16x24 feet outside, large for a one room cabin in its day. The floor was roughly level but rolled to one end, and the board ceiling, which hid the ceiling rafters, was a little more than seven feet high.

I wrote down my impressions in a notebook I'd brought with me, then headed home. There was little else I could do or tell until we were further along.

When I stepped into the kitchen, I could hear Liz laughing on the phone. "What's the name of the guys who sold you that cabin?" she whispered over the mouthpiece, I told her, she nodded "Yup" into the phone and laughed some more.

In a couple of minutes she got off. "That was Paulene, and those guys you're working with, Carl and Gerald, are her sons," she smiled. "They're tickled pink to know it's you who bought the cabin."

Paulene is an older woman, native of this area. Liz first met her when she responded to an ad for hand-made quilts on the local radio station's "Listener's Exchange." Liz bought several, and they became friends as she sold many more quilts for Paulene to our friends and family.

Next day when I went to work on the cabin I took along my son Marcus to help. When we arrived, well it was like old home day. Carl and Gerald treated us like lifelong buddies. They told me how much "Mama" thought of Liz and I conveyed how much Liz liked her, as well. From then on, we worked side by side as friends, helping each other as best we could every step of the way.

It was a good thing, too, for taking down that house, and the cabin proved to be much more work than any of us had envisioned. About the second or third day after we started working on it an August heat wave rolled in and wouldn't let go. I remember meeting them there early one Saturday morning and the temperature was already pushing the mid 80s. By mid-day it chased 100, and those temps hung around for weeks.

Eventually, Carl explained that in exchange for taking down the house he and Gerald were entitled to all the lumber, tin and whatever else they could salvage. The owner only wanted a single square poplar beam which had been stored in the attic loft. It was massive—20 x24 inches around and 20 feet long. It also bore the distinctive signs of having been mortised, a building technique we later discovered had been used in the rooms built around the cabin.

Gerald, Carl and his son-in law worked as a team, and Marcus worked with me. For some of the larger jobs, like taking down sections of the roof, it took all five of us, working together.

Dismantling the house required that we work in exactly the opposite order in which it had been built. The cabin would be the last thing we'd get to as it was the first part built. For the next few weeks, it remained isolated within the framework of the house slowly disappearing around it.

By the end of the first week the halls and floors throughout the house were filled with drywall, debris and scraps of boards. There was so much of it, and we had only started, it was difficult to walk.

So we removed all the weatherboard siding from the side of the house, all the way up to the second floor. Then we just threw the boards outside as we removed them. We weren't particularly careful but few boards broke. Soon tall stacks of boards and junk permeated the yard.

Most of the boards were 1x6 mill cut poplar or chestnut, and the floors were usually linoleum on tongue and groove pine over the original poplar or chestnut plank floor. Many of the rooms had a "beaded" ceiling of pine commonly used in homes all over this area through the 1930s.

The first rooms completely dismantled were the upper bedrooms in the addition to the house. Everything was framed using true 2x6 and 2x4 lumber, held in place with large cut nails. The beaded ceilings had been painted white repeatedly. When we took some down we found the original electrical wiring. Some had apparently still been in use the last time anyone lived in the house, less than 10 years ago.

It was probably an example of the earliest electrical wiring used in the area, and seemed quite curious to us. It consisted of a pair of wires, one insulated and one not, which ran parallel to each other over and through many of the ceiling rafters. Wherever the bare wire passed through the rafter it ran through a ceramic tube, about six inches long, to prevent contact with the wood.

After that first full week we had reduced the house to its framework. You could now get a good sense of the cabin within but much work remained before we could begin to tackle that room. Now, though, with sections of its walls uncovered you could recognize the log cabin there. I was happy to see *most* of the logs were fully intact, and well preserved.

Except at the end near the chimney, facing west, for several reasons. In the years that the roof had leaked onto the logs it had accelerated the decay which had already begun around the chimney. If there's rot or decay on any of the old cabins I've seen, it's usually near the chimney, Over time the stone work directs rainfall onto the logs. When there's no way for the logs to shed that water, and they remain damp, and dark, it's only a matter of time before they start succumbing to the elements.

That process had likely begun a century ago, before the cabin was covered up, and somewhat protected. When the roof was torn off, though, the rain and snow fell straight down and inside the walls. There it gathered and lingered in the dark spaces, exposing those logs to constant moisture for several years. There was little to be salvaged from that end of the cabin.

Until then, I thought about rebuilding the cabin as a separate building next to my office. I had planned to add a "dog trot," or covered walkway between the cabins. I realized I would have to revise my plans.

From what I could see of the outside of the cabin the rest of it was in tact. I wouldn't fully know for another week, until the inside walls were completely uncovered, just what I had to work with.

The immediate challenge was dismantling the roof and what remained of the frame built around the cabin. We started on the top floor at the far end, trying to take down the roof rafters one pair at a time. First, we had to remove the 1-inch thick boards which had been used for lathe to support the tin. Unlike modern tin roofs, composed of sheets of tin, this roof was actually one large sheet of tin, made up of smaller sections crimped and soldered together.

Taking it apart, and getting it down was exhausting work in the hot summer sun. The size and number of cut nails we had to remove, rafter by rafter, really slowed our progress. Also, we had narrowly escaped injury when one set of rafters suddenly collapsed where we were working, we were more cautious and deliberate about the work.

Carl finally decided this was taking too much time. To speed things along, he wanted to tear the roof and rafters free with his four-wheel drive Chevy truck. He had some of the same discarded utility cable I had used to haul logs out of the woods, probably from the same source. We looped and tied it through the rafters and to the bumper of his truck. Carl put the truck in low gear and started driving away from the house.

The wheels spun but the rafters didn't give an inch.

He tried backing up a little and racing forward. This time the truck lurched, the cable snapped but the rafters stood, defiant.

Next I went and retrieved my logging chains, and we combined them with several lengths of cable. Carl backed his truck almost into the building, and surged forward. The truck jerked to a sudden stop again, but the cable, chains and rafters held.

For a couple of minutes we debated whether or not another try would jerk his truck axle loose. Finally, Carl backed up to the house, but this time he revved the engine before popping the truck into gear. The tires screeched, mud flew, the truck roared forward and the rafters followed, the last two having broken midway under the strain.

"They sure don't build them like that no more," Carl announced with a satisfied grin as he stepped out of the cab.

We went right back to work. By the end of the day we also had the ceiling boards down and were starting to dismantle the walls for those rooms. In the weeks that followed as we worked we were continually impressed with how well built the old house was, the quality of the lumber and the size of the cut nails.

When we finally worked our way down to floor level, though, we found all the framing lumber was mortised into holes precisely cut into large hewn timbers at the base of each wall. We could only marvel at the amount of work and care that had gone into building that addition.

As we worked, here and there we came along the odd treasure or trinket. Gerald found an old pocket knife in one wall, and Carl lucked onto a length of old gold chain which a rat had carried off to a nest in a corner of the ceiling. There were also pieces of toys, a shuttle and spool from a loom, a hand made paint bush, pencils and other odds an ends. A set of finely painted glass slides, used to project pictures on the walls with an oil lamp, suggested the family enjoyed an affluent life at the turn of the century while others in the area were scraping by.

The more we removed the closer we got to taking down the cabin. Soon it was completely uncovered, inside and out. Apart from the end by the chimney the rest of it was pretty sound. Around the back doorway two logs had extensive areas of rot from a combination of exposure to the elements and insects, probably termites. And somebody had painted their name on one log decades ago. Nevertheless, there were more than enough logs for the reconstruction I planned.

In preparation for dismantling the cabin, still a few days away, I took some time to mark the logs so I could assemble them in the right order. I tacked a piece of plastic fiber about 2-inches by 4 inches to the end of each log. Then, using bright red paint, I painted a number on each.

I designated my corners A, B, C and D, and started numbering the notched logs from the bottom log up A 1, A2, A3, etc. By the combination of letters on each I could later tell which wall the logs belonged to. B-C, was the back wall, for instance, and A-B the wall opposite the chimney. Sections of logs which didn't include a notch were marked by the wall and location, A-win1-door-D for the log section which ran between the first window and door on the entrance wall, for example.

Happily, when we uncovered the cabin we found the interior and exterior walls had survived better than the rest of the house

We were almost ready but now our work was significantly slowed by the amount of lumber and debris around us. It was piled outside and in every pathway throughout he house. You can't possibly imagine how much goes into a house until you take one down and try and haul it off

Even though Carl carried a huge truckload of usable lumber home every day we were overwhelmed with building material and junk. We had to break piles into smaller piles just so we could move around.

Sometime that week Gerald mentioned a friend had told him they had sold the cabin to me too cheaply, that logs from old cabins like that were bringing thousands of dollars in Pigeon Forge, Tennessee.
I didn't know what to say, as I'd spent all I could afford.
Fortunately Carl stepped up . "Well, we made a deal with Mike, and we stand by it," he said. "Besides, look at all the help he's given us, and his boy. That's worth something." It was never mentioned again.

And a few days later we finally started attacking the cabin. We began on a Friday afternoon with Carl climbing up on the roof and tossing down the long sheets of tin while Gerald, Marcus and I worked below in the loft.
Beneath the tin roof was a layer of lathe, rough cut slabs of logs. In one corner toward the back of the cabin a small section of "shakes" remained, hand made wooden shingles from the original roof, were tacked to it. There were about five or six, roughly a foot wide and as much as two feet long. Friday night, as the sun went down I pried a few loose as souvenirs.

Saturday morning we actually got down to dismantling the cabin walls. In the living room tongue and groove boards had been put up long ago to cover the ceiling rafters. The rafters themselves were massive oak true 2x8s, 16 feet long. They bore the same distinctive up and down saw marks as the rafters used in the loft of our original cabin at home. It was like a signature that said to me they had been cut at the same mill, with the same blade.
With the roof and rafters out of the way, we still had to remove the loft floor before we could start taking down the logs. As we pried up the first of these floorboards we were startled to see they were held in

place with hand-carved wooden pegs. No nails, just these carved poplar pegs. Either that floor was put down so early there was no source of nails, or the early owners had been too poor to afford cut nails when they first added the loft.

That hinted something about the age of the cabin. Still we really didn't realize how old that cabin it was, or how early a specimen, until we actually started to remove the logs. Carl, Gerald and I discussed for some time the best way to dismantle the walls. I had thought about setting up tripods and lowering the logs one by one, but they stressed there wasn't enough time, that the owner was already on them because the job was taking so long.

As an alternative, we piled lumber against the outside walls so we could push the logs off the wall for just a short fall before rolling them off the piles onto the ground.

Here, at last, the real work began. Some of those logs must have weighed a ton, literally. The cabin wasn't built from the expected mix of lightweight poplar and chestnut. Rather, it had been built of a good and representative selection of the many different types of trees once native to the area: poplar and chestnut, yes, but also oak, hickory, walnut, locust and ash.

It was all we could do, working as a group, to lift some of those logs high enough to roll them out of the way. We grunted, and snorted, dragged, pushed and pulled. Think of what a piece of red oak or hickory, 24 feet long, 12 inches wide and eight inches thick, must weigh!

Yet one by one, through the smothering heat of that day, we brought the logs down and moved them out of the way. With many, it took all our effort to get the logs to the flatbed trailer nearby, and slide them in place.

Based on observations as the cabin came down, I reached two conclusions. First, that this had to be a very early if not one of the earliest cabins around. I had some idea of that from the wooden nails and the hand carved pegs. But now seeing the selection of trees used, it was apparent whoever built this cabin had done so without much knowledge of the local woods. Later builders preferred the lighter poplar and chestnut logs. This cabin had been put up using whatever trees were on site. That seemed the only criteria used in their selection.

I also reasoned that whoever put this cabin up had to have additional help. We—five of us—struggled to move some of these logs nearly 200 years after they had been cut; I can only imagine how much they weighed when they were green!

That help may have come from a mule team, work horses, oxen, even slaves. Some of the earliest settlers in this area brought slaves with them, and slave labor may have made a crucial contribution to the construction of this cabin.

We were, for that weekend, slaves to the exhaustive challenge of dismantling the work done long ago, and moving the logs out of the way. The day's heat forced us to pause many times for a catch of breath and drink of water. Moving the logs and their mud chinking stirred dust which clung to us everywhere we sweated, covering us with a brown grime.

At times Gerald was so hot and tired he could barely talk or move. We all felt overwhelmed by the labor but forced ourselves to continue. Had we sat down too long I doubt we could have continued the work.

I had borrowed a flatbed trailer from my friend Drew for hauling the 24-foot logs, and Carl hauled some of the 16-foot logs from the shorter walls in his truck. Since my Chevy Luv pick-up didn't have a hitch, I had to pull that trailer with our family car, a 1978 Toyota Corona sedan. Because of the weight of the logs I could only pull a few at a time. In fact, the weight of the logs was so great that it pulled down the rear and raised the front end of the car into the air.

When I hit a bump the front wheels would suddenly lift off the ground for a few seconds. After a couple of trips I learned to crawl along the road and avoid the rough spots. Fortunately I only had to pull the trailer less than two miles.

We used the car and truck to haul the logs because the owner was eager to have the mess cleaned up as quickly as possible. Carl asked if we could stack the logs in my yard temporarily, promising to return with his truck later to haul them up the hill to where I planned to resurrect the cabin.

By the end of that long long day we had all the logs down and scattered about the yard, but we had only hauled a couple of loads back to my house. The boys couldn't work Sunday so I talked friend Drew into helping Marcus and I move the rest of the logs. After several slow trips, there were piles of logs, plus siding and lumber they'd let me have, stacked all around our yard. It gave people something to slow down and wonder about for the rest of the week.

When I recalled what the cabin had looked like standing, the piles in my yard just didn't seem to add up to as much. It looked more like a jumbled mess of logs and lumber, dusty and dirty, smelling with the mustiness of an old house. I tried rearranging some of the logs into neater piles late Sunday but found I simply couldn't move a few.

Either they were so heavy or I was too tired from the weekend's work.

The following afternoon, Monday, when Carl and Gerald didn't show I figured they were just as tired. I piddled around, by myself, adding to the piles of trash and lumber around the cabin site. Then, I started pulling up some of the floorboards in what remained of the cabin. Beneath the top layer I found rougher planks which, like the loft floor, were held in place with carved wooden pegs. It kept me occupied until darkness arrived.

I didn't feel much like working the following day, Tuesday, but headed over to the cabin just before sunset to see if the boys had been there. Marcus was with me and he and I were moving more lumber when I heard a car approach. I didn't even bother looking up, assuming it was Carl or Gerald.

Soon after I heard the car door shut, though, I recall a woman's voice, obviously distraught. "Oh my God…Who is responsible for this…Who did this?" she moaned. It wasn't the best introduction.

In a few seconds a woman who identified herself as a descendant of the family which built the cabin was confronting and accusing me. She held both hands to her face as she whined "Was it you?...Who told you could do this...How could you do this?"

I gave her a look which should have said "What the hell are you talking about?" but she continued her high drama.

"Oh, I feel like I've been raped!" she gasped at one point, then allowed "I've already talked to my lawyer to see what I can do....You can't do this...I've got to have my cabin...You have no right to do this"

When she finally calmed down enough to explain herself, she told me she was a descendant of the original owners. She had dreamed of one day restoring the cabin as it had originally been built, as some sort of monument to her family. She had seen us working on the house while driving by the previous week, and had spent the days since trying to find ways to stop us. Apparently she learned she really had no valid claim on the place.

I explained myself and how I had come by it, for $200.

"$200!!!," she stammered. "That's not right! I've got to have it back."

It didn't matter to her that the place had been sold by one of her relatives years earlier, or that the cabin had stood neglected for years, or that I had purchased from someone operating with the present owner's consent. She considered it rightfully hers. She even produced a copy of a will which she said indicated the farm should have been kept in the family indefinitely.

Well then why wasn't it left to her, I reasoned, rather than to the person who sold it off.

And although I could sympathize with her at one level something about her timing perturbed me. "I can understand how you feel, but I'd feel a whole lot different if you had come to me before we did all the work in taking this cabin down," I told her.

"You should have said something as soon as you saw someone working on the place."

She wouldn't hear me, didn't want to. She just insisted over and over that she had to have the cabin, that it must remain in her family, that this was a lifelong dream of hers, that what we had done, and what I planned to do in rebuilding the cabin, just wasn't right.

Finally she got in her car drove off. But the next day she called and harangued me while I was trying to work, as well.

By then I had a pretty good idea where I stood. After she left I had rushed home to call Carl and ask him what he knew about it. He was caught as much off guard as I was, and had been preoccupied since Sunday. Gerald, it turned out, had had a heart attack, or something approaching it, after our exhaustive work on Saturday, and was in the hospital.

But Carl put through a call to the owner who wasted few words. "He told me he bought the place, and said I could do with it what I want and you paid me for the logs so they are yours," Carl relayed in a call to me later that night. He also advised that I "Tell that woman she's crazy."

The next day when she called she ran on about the cabin and how much it meant to her and her family. I listened, and thought about all she said. But I pointed out her family had relinquished all claims when they sold the place. She was insistent though, she had to have the logs, even offered me the $200 I had paid for them.

"That wouldn't begin to cover the work we've done, and I'm not selling them anyway," I countered.

I left it at that, but meditated over the situation from every angle for the next few days. The next time she called, I made her what I considered a fair offer. Looking back I guess it was a mistake to even try as it only prolonged a difficult situation for both of us.

"Here's what I am willing to do," I told her. "That cabin measured 16 x24 and was 1–1/2 half stories high. You find me a cabin of comparable size or the logs to build one, delivered to my home, and I'll let you have this one."

She agreed to that. But Saturday morning, when I went by the site to help Carl finishup she was there, with a few other people. A wrecker was tearing out what little remained of the cabin, the original log floor joists.

Apparently at her direction, the tow truck driver had tied his tow cable to the hewn floor sills and log floor rafters. Then he just drove his truck to pull apart what was left of the cabin.

There were deep ruts in the field where he had dragged these logs and pulled them onto a flatbed trailer. The entourage watched until the driver finished his work, then drove off.

I wasn't sure what to do.

Next it was I who called to remind her she had agreed to provide me with a comparable cabin. She needed a little more time, she said. More than once over the next week Carl questioned my sanity for even trying in any way to accommodate her.

She finally came up with cabin to trade, one I knew well but she had never seen. "That's barely a corn crib," I told her. "And it's in terrible shape."

Another week went by before we spoke again. She still hadn't come up with anything acceptable. During the course of that week I had also paid a visit to another of the family homes the old man had tried to maintain while alive. I couldn't see this one for the weeds in the yard, towering as high as the roof. I waded through them for a look inside, and found windows broken, doors ajar and the bottles and ashes which indicated it was a regular stop for late night partyers.

This time when she called I cut her off. "Look, this has dragged on too long. I've tried to be fair but you've really not come up with anything that's acceptable."

She pressed further and I got fed up. "If you cared so much about the damn place why didn't you patch the roof when it was torn off, rather than let it leak onto the logs all those years and ruin one wall.

"And besides," I continued, mounting my high horse, "if you're so concerned about family heritage, how about taking an interest in that other homesite. The weeds are overgrown, and the windows are broken, it could sure use some special care." She had no response.

Finally, I told her "I want to move on. I'm going to build the cabin as I planned. It took a lot of work and I bought it fair and square. When it's up, you or any member of your family can come see it, come stay in it for all I care, but I want to move on."

She finally and reluctantly relinquished, and told me where I could find the logs they had taken.

Again I got Drew to help me, and we retrieved the logs and sills where they had been stacked in a driveway. It was the middle of the day but I was still worried she was going to appear and make a scene or call the cops and accuse us of stealing. Our mission was accomplished, without incident. I was never bothered by her again, but I've been told she still feels I "got her logs."

The cabin sat stacked in my yard for the weeks while this drama played out. Soon as I called Carl, though, he was there with his truck to help me load them onto the trailer and pull them up the hill. Gerald was still recuperating and Marcus wasn't home so he and I worked together for a couple of hours loading and unloading the logs. If his truck hadn't been a four wheel drive we never could have gotten them up this steep hill.

"Well I'm glad that's finally over," he announced as he wiped his brow after we rolled the last of the logs onto a pile.

I could not have agreed more—even though half the work still lay ahead of me.

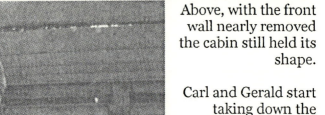

Above, with the front wall nearly removed the cabin still held its shape.

Carl and Gerald start taking down the ceiling rafters in the cabin room as Marcus looks on.

After we were through, it was hard to imagine the cabin in this pile of logs in my front yard

Trials and Tribulations With An Unexpected Find

I knew then—it was now October—I would not have the time to put the cabin up until the following spring, at the earliest. My first worry was stacking and storing the logs where they could remain for some time. Because the logs were so heavy I wanted to move them as little as possible. Since most from the end where the chimney were useless, I decided to make three piles, one for each wall I planned to build.

I made platforms for stacking the logs from sections of the damaged logs set on concrete blocks. Then I sorted the piles based on the coding I used to mark each wall and each log. To move them around, I again employed the "log wheels" which had served me so well on my earlier projects.

By a stroke of luck my son Marcus was working at a nursery at the time and they were re-doing all the plastic greenhouses. He was able to furnish me all the discarded plastic tarps I needed to cover each pile.

It seemed an ideal solution. After a few days, however, I noticed there was a lot of condensation on the inside of these covers. This moisture actually created a more serious problem, as I discovered when I removed the tarps. Long ago, in their initial attempts to hide the logs, the folks who lived in the cabin had covered them by pasting newspapers over the walls. Their paste was most likely a simple mix of flour and water.

Under the tarps, the constant dampness created the right conditions for mold to grow on the flour paste. In some places this mold had really taken hold, and given rise to another yellow jelly like fungus. To make matters worse, the extended exposure to heat and humidity was also promoting the resumption of some decay which had started long ago.

I had to undo the piles and roll the logs around to determine the full extent of it, and to scrape away whatever mold and newspaper I could find.

Thereafter I decided it best to keep the logs uncovered as much as possible. At the slightest hint of rain, I would roll out the plastic and cover them again, only to uncover them the next day. Through the bitter part of winter, and subfreezing temps, I kept them covered all the time.

During the warmer months, though, this set-up meant a lot of racing around at the first rumble of thunder to get the piles covered. On windy days, it could be a major chore just trying to weigh down the massive tarps so they would not scatter or tear.

Now that I was alert to potential damage, I decided it was important that I check on the condition of all the logs every couple of months. That meant moving all the logs and rearranging the piles, an exhaustive but necessary afternoon's project.

I had to live with this arrangement for the almost a year and a half, as I was still finishing up my office at the time. Despite my best and consistent efforts, I could not adequately protect all the logs. A few had enough decay started to entice beetles which chomped their way through the wood. By the time I found the piles of sawdust it was too late.

Worse, and more vexing were the termite trails I found one time while rolling the logs for the long wall opposite the door. The trails led to two logs, both oak, but I undid the entire pile and examined every log in search of any hint of infestation. Those were the only two fortunately, and I had discovered this early enough to save them.

I've always tried to avoid using pesticides as much as possible but felt I had no choice but treat those logs for termites, and spray the entire piles to protect against any further damage. Fortunately these efforts worked and I found no further signs of the critters.

After another winter, I was finally ready to begin rebuilding the cabin, or something closely resembling it. By then our sixth child, Peter Zechariah, had been born. With four teens and two tykes running around, our house was now quite crowded. I could justify the expense involved in building another cabin by reasoning we could use it as bunk house for our two older sons. Marcus and Luke were both in their late teens, and I reasoned that living quarters of their own would be a good start toward independence.

Since most logs at the chimney end of the old cabin were useless, I planned to rebuild the three salvageable walls onto the back of my office cabin. This would roughly recreate the large room with interior dimensions of 15 x24. Upstairs I planned a loft with two sleeping areas roughly 8x10, enough for a bed and a dresser.

The large downstairs room would look just like the cabin we took down, with the same entrance doorway and windows. On the opposite wall, though I planned a window instead of the original door.

Surveying all usable logs against my plans, I had enough to build three sides, but only nine feet high on each wall. For their bedrooms, I would have to use modern framing construction; I was not about to get out my broadaxe and adze again.

I also found that many of the notches at the corners of the long logs where they had joined the chimney wall logs were damaged or decayed. This seemed the result of the exposure suffered during the years rain and snow had slipped into the walls. If I wanted the full sized room, I would have to cut the logs' length and improvise

Now that I knew what I had to work with the rebuilding project finally could begin.

I started, as with other cabin projects, by building the floor as a deck. It stretched away from the outside of the back wall of my office cabin. This time, however, I built the floor the full size of the outside walls, as the bottom or sill logs were not usable. In their place I was going to use squared logs which had been used as sills in another section of the old house. Since they were square, I planned to set them on the floor for the long walls, and reconstruct the original cabin from that first course up.

And, as before I used lumber cut at a local saw mill, starting with 6x6 poplar sills, topped with 2x6 poplar floor rafters. For the sub-floor

I again resorted to cull lumber, one inch boards of varying widths and types, including oak, cherry, walnut and sassafras.

With this floor down, I had to figure out how to secure the logs at the ends where I had to cut back the notches. If I were to notch them into the existing walls of my office cabin I would end up with a much smaller room, so I compromised. I cut the rotted notches from the end of each log, reducing the total length to 2o feet. For four feet between my office and the old cabin I planned to frame the wall with 2x4s, and cover them with siding for walls with a total length of 24 feet.

I still needed a way to stabilize the logs at that end, and to tie the framed section in with the logs. For this, I purchased two pieces of four inch angle iron, 10 feet long. I had a machine shop drill half-inch holes every six inches along both sides of the angle iron. Then, I set the angle iron in concrete, Once the concrete set I could bolt the logs to the iron at the ends where I had removed the notches.

Angle iron in place, subfloor down, the cabin reconstruction moved into high gear. Marcus who had helped take it down, was there every step of the way to put the cabin back together. Often we were joined by Sean then 13, and occasionally Luke, 17.

Before we started building, we re-sorted the log piles for each wall so that the top logs were at the bottom of the pile. This way we could retrieve the logs as needed, without any additional work involved in sorting through the piles.

For the back wall, which originally had a doorway, I planned a window. There we substituted full-length logs from the old cabin's loft for the half logs which had framed the doorway.

After more than a year of labor, anticipation and planning, putting the cabin back together was surprisingly easy—at least for the first few courses. It was almost like working with Lincoln Logs, on a grand scale!

Since the lower section of the walls, except for the back wall, were broken up by windows and doors, we were working with sections of logs. That made it easy to maneuver the logs into place, but the weight of some of those oak and hickory logs was still staggering

The only notches we had to trim or re-cut were on the substitute logs we were using from the loft; everything else fit like a glove. The coded marks I had tacked to the notches made it easy to put the cabin together just as we had taken it down. Still, I continually compared our progress against the pictures I had taken of the cabin to be sure.

We had to build tripods to raise the full length logs and some of the heavier sections with block and tackle. By now, I'd done it often enough that it was a snap building them. Within a few days, we had rebuilt the walls to the tops of the windows and doors.

I was worried though, about the size and weight of one of the remaining logs. It was massive, almost a full 14 inches wide across its face, and had been used to support the ceiling rafters in the old house. Even though it was poplar it weighed a ton! The tripods creaked under its weight, and it took two of us pulling on the lead to get it started off the ground. Soon we had it up, and pulled it with ropes to set it in place.

Even I was impressed when we had the opposite log up and prepared to set the 16 foot 6x6 polar beams in place to support the upper rooms. Our reconstructed cabin was such a close match to the original we could use some notches which had supported the ceiling rafters in the original cabin with only minor trimming!

We set five of these 6x6s in place, every two feet apart and laid down the plywood sub-floor for the loft rooms.

Then, we set a final course of logs over these beams. Next, I went to work with a level, chisel and hammer to level off the top logs to serve as plates for the roof, and as a base for the framed walls which would enclose the loft bedrooms. From there, the project involved standard carpentry procedures for the framing the roof, and building the interior and exterior walls for the loft.

In fact, at that point I was satisfied most of the work was complete, and the project could continue without incident.

The entire time I labored on these log cabin projects I was also working full time as a freelance writer, and trying to reclaim some vestige of our "farm" from the thick woods which had overtaken it over the last half century. Much of this involved cutting trees and piling brush, which I later burned off.

One crisp fall morning, after we'd put up the logs and were about to start on the framing, I decided was the perfect day to burn off some of these brush piles.

The sky was clear and the air calm. I called the forestry department for a burning permit and set to work igniting several of these piles. One of the smaller ones set 150 feet down the hill, at a safe distance from my office, or so I thought. I tended the fires as I worked that morning, repeatedly leaving my desk to check on them, and rake the outer debris into the center of the fire.

By mid-day the fires appeared to have burned themselves out. The air was still calm when I had to run into town to pick up Matt, our second youngest, from nursery school while Liz stayed at home tending the baby. It's a 20 minute round trip, if that.

On the way back from town, as I started down our road, I saw a hint of smoke two miles ahead and thought nothing of it. At that time of year people are always burning off brush piles.

As I made the last turn toward our place, though, I was absolutely horrified to see our entire hillside cloaked in smoke, the silhouette of the cabins standing starkly against it. The office is on fire!!! I thought.

A thousand thoughts raced through my mind. I had not connected a water line to the hill yet and it would take too long to hook one up now. I feared time was running out, if it wasn't already too late.

Rousing Liz, with a shout that "The hill's burning!" she loaded the baby into the car while I filled any container that would hold water and crammed them into the trunk. Then, we roared up the hill through the smoke and fire rimming the edge of the road.

In the few minutes I was gone a wind had come up at the head of an approaching front. It had been enough to fan the embers below my office into a blaze and then scatter them around the hillside. It ignited the line of fire which now stretched along and up the hill. It had already seared through the grape arbors and orchard and was progressing toward the office.

The wind was still up, now fanning the spreading fire in every direction. It didn't take a minute to empty the containers in an attempt to create a damp barrier between the flames and the office. Liz sped down the hill through thickening smoke, with the kids still in the car, to refill the buckets while I tried to contain the fire. At that point it seemed the old trailer which had once served as my office, and still held much of my background material and historical research, was most threatened.

Grabbing the rake and shovel kept on the hill, I started scraping the ground of leaves and other debris to create a firebreak. The smoke from the burning leaves was overwhelming at times, but to pause would be to surrender to the flames.

Liz continued to make trips up and down the hill with our bucket brigade while I worked to save the trailer. During one of her trips to the house the school bus brought home three more of our children, Sarah, Luke and Sean. We immediately put them to work raking a break on the far ends of the fire while Liz and I concentrated on the rapidly progressing fire near the office.

No sooner had we succeeded saving the trailer than I recognized another loss. I had kept all the lumber for framing the addition in a neat pile, which I had covered with the plastic tarps I had once used to protect the logs. While we were focused on trying to save the trailer the flames had pushed near enough to ignite the edges of the plastic. It dripped onto the wood as it rapidly burned, setting fire to the inner portions of the stack.

The few available buckets of water thrown on the pile could not douse that fire. We had to watch as hundreds of dollars worth of lumber roared ablaze. The flames and heat were intense but Liz and I got as close to them as we could stand, covering our faces with our shirts, while we tried to scrape a break in the dirt to keep this fire from spreading further.

"Your office!" she yelled suddenly, as I looked to see the fire had advanced across the road, even where we had thrown water. It was now pushing along the path toward the wooden stairs. We jumped to work there, stamping it out and scraping another break in the dirt

Only then, with the help of the kids racing around the perimeter of the blaze, were we able to gradually gain control over the fire. It still took another hour of tracking back and forth around the edge of the charred patch in our woods before the fire was finally and definitively extinguished.

Exhausted, raspy throats, covered in smoke, we surveyed the damage. 'It got within 1o feet of your office," Liz pointed out, showing me the blackened fingers of burnt grass which seemed to be clamoring for my rounded log cabin, and the cabin walls we had just rebuilt behind it.

It was another two weeks before the saw mill could cut and deliver the poplar lumber so we could finish the work on the cabin. I was quite grateful, and gave repeated prayers of thanks, that we'd lost so little.

I'll never leave a fire unattended again until the last ember is out.

After that, completion of the project seemed almost anti-climatic. The boys all helped as best they could but this experience had exhausted me. When at last we finished the drywall and painting, Marcus and Luke moved into their new rooms on Thanksgiving Day.

The loft rooms are a little cramped but it's a nice big room downstairs, with a kitchen area, and ample seating area for entertaining their friends. The kids keep the place a mess, but that's their business, I guess, or so Liz keeps telling me.

We still debate whether the cabin is haunted, and if there are such things as ghosts whether they would follow a home, or what made up the home, to a new location.

Sometimes it just seems like something is there, some sort of presence. We all agree on that. And Marcus swears one night as he was falling off to sleep something came into his room and sat on the edge of his bed, but he was too scared to look and see if it were a cat, or more.

If it was a ghost, I bet it was just some weary spirit trying to catch a moment's rest from all the work that went into this place.

It was enough to cure me of log cabins.

Awaiting some trim, paint and a porch roof, the reconstructed cabin looks much like it may have when built.

2

Methods for Taking Down and Rebuilding A Log Cabin

FINDING A SALVAGEABLE CABIN

Sheltered from wind and weather, log cabins can be surprisingly enduring. In rural areas where they once dominated you'll still find survivors you can take down and rebuild standing alone, or hidden within other structures.

Look in the same places where you might find cabins to restore(see Part I, Chapter 2). In fact, it will probably be easier to find a cabin, or logs, you can dismantle and re-use, than locating a few acres you can buy with a standing cabin to restore. If you're heading out to find a cabin, review the clues to look for in Chapter 2 of Part I as well.

DOCUMENT BEFORE DISMANTLING

Rebuilding a log cabin can be as easy as assembling a model kit, almost—provided you take the time to meticulously document the cabin as originally built. With detailed notes, pictures and carefully marked diagrams, you'll have a guide to rebuilding the cabin.

So, the real secret to rebuilding a log cabin successfully is in all you do before you even begin dismantling the old cabin walls. To recreate that cabin in a new location, you've got to know just how it was put together in the old.

Everything about a finished log cabin works together: the shape of each notch, the length of the logs, the run of rafters are all determined by their location in the original structure. Once you've got piles of logs stacked on the ground they all look alike.

If you don't know which goes where you'll spend needless hours and effort trying to put the cabin back together, and may never get it right.

Document what you start with, and you'll have a realistic idea of what you should end up with.

How Will You Use It?

If you're not sure how you may eventually use the logs, you'll appreciate the effort that goes into documenting the cabin and marking the logs if you ever do decide to rebuild it.

When you plan to dismantle a cabin as a source for logs you will to cut and use for an entirely new structure, there's no need to worry about where each log fit in the order of the wall.

But if you intend to rebuild the entire cabin or only a portion of it, you'll need all the details you can gather about how the place was first put together when it comes time to rebuild the cabin.

Take Plenty Of Pictures

There's a story in each log cabin, and a good part of it can be told with pictures. As soon as you find a cabin to take down, start taking pictures, and don't stop until the last log is stacked.

Initially, take pictures of the cabin as it sits, or sits inside some other structure. When the cabin is a hidden within a home, these first pictures give you a record of what you started with. Then, as you remove the surrounding structure and get down to the original cabin walls, the pictures you take will serve as a reference for how the rebuilt cabin should look.

Once you've isolated the cabin as it was originally built, document all structural details. You'll need photos of each entire wall, inside and out; shots of the corners showing how the logs are notched together; pictures of the construction of each door and window frame, and where the floor and ceiling rafters intersect the logs. If you plan on a full "historic" restoration, including the loft and roof, you'll want pictures detailing these features as well.

You'll get the best results taking pictures of the outside of the cabin on overcast days when there won't be a lot of shadows. For interior shots you'll need to use a flash, and artificial light if it is available. Including a yardstick or tape measure in your pictures, as a reference, will allow you to figure out the actual size of what's captured in each photo.

Document the cabin with a film or digital camera. Digital wasn't available when I did my work but I think it offers a couple of advantages over film. With digital, you can review your images on the spot and re-shoot any as needed. Since you can easily load them into your computer, and enlarge or crop them at will, it's easier to glean the information you need from the pictures.

If a video camera is available to you, I'd suggest you also tape the cabin as it stands, as an additional reference. Don't rely on video exclusively, though, as it won't give you as accurate a perspective as straight on photographs.

KEEP COPIOUS NOTES

Your photographs should be backed with detailed notes about the structure.

Keep a journal on the project, and add to your notes as the work progresses. Begin with your observations of the cabin as you find it, and record your impressions as you learn more about the cabin, and the house or barn which was built up around it. As you remove the surrounding structures you're turning back pages in its history. Any addition, as well as the type of materials used, tells something about former inhabitants and the era in which they lived.

Record any artifacts you discover as you work toward the cabin, for they too tell part of the story. A horse hair paint brush, the shuttle of an old loom, or a smooth coin fallen through cracks of the floor are echoes of its past. You may even be able to read some of its story, literally, from the newspapers and catalog pages pasted to its interior walls. If you can't save some of these pages themselves, be sure to record any dates you find. The earliest date you find will help estimate the age of the cabin.

Once you get down to the original cabin, tag the logs and all building materials(see next section). Then take and record detailed measurements of the entire structure. Measure both the interior and exterior dimensions of the length and width of the cabin; make precise measurements of the openings for doors, windows, and the chimney—anything which affects the shape and construction of the building.

If there's a stairway to a loft, note where it is affixed to the wall, how far out it sits from each wall, and the size and location of the opening and/or door.

Assess The Structure

While taking these measurements, assess everything about the structure. Examine the condition of all logs and the notched corners. Note any logs, corners or sections of walls which show signs of progressed damage.

Also, closely inspect the chimney and hearth if still standing. Look for signs of loose bricks or blocks or indications that what remains of the chimney is leaning on the cabin for support.

If there is any doubt the chimney can stand on its own you'll need to take it down before you dismantle the cabin. Otherwise you'll be exposing yourself to constant risk as you work on the logs.

Isolate The Cabin

Before you begin work on the cabin, completely isolate it within or from the surrounding structure. This may require that you completely remove additions added to the cabin after it was built, or that you physically separate the cabin within the rest of the building.

You may need to prop up sections of the roof or ceiling with new support posts, or brace walls which had been tacked on to the cabin. Take the time to make sure any part of the structure which once rested on or against the cabin is adequately supported so it can stand on its own before you start removing the cabin.

This will ensure your safety, and make it easier to take down the logs.

Removing The Chimney

Keep safety first when dismantling a chimney. Wear protective gear, including a helmet, and never try to remove large stones overhead. There's always the danger of an old chimney collapsing if it's leaning against the cabin for support. Even leaning a ladder against a weakened chimney can put enough pressure on it to bring it down.

Before you begin dismantling a chimney assess its condition from all sides, and carefully plan how to proceed.

If you plan a precise reproduction of the cabin as built, number each of your stones on the top and inside as you take them down, course by course, and describe the layers and construction of the chimney in your notes. Even if you don't plan to rebuild the entire chimney, it's advisable that you mark the blocks which make up the hearth. Many of these may have been specially cut, and knowing what goes where will make it easier if you ever want to recreate that space.

When you aren't concerned with preserving the rocks, the easiest way to quickly bring down a the chimney is to wrap a heavy chain around it about halfway up and pull it over and away from the building with a heavy duty truck or tractor. If you are going to do this, make sure you have enough chains to put the vehicle safely beyond reach of any damage from the falling stones.

If that's not an option, and the chimney can not stand on its own, you'll need to dismantle it, course by course, from the top down. This requires scaffolding or a pair of ladders set up so they do not rest on the chimney for support.

Then, with hammer and chisel chip away the mortar holding each block in place. Work slowly and carefully. Lift the rocks from above and move them out of the way. If you can't devise some means for safely lowering the blocks toss them away from the structure to the ground. Always check to make sure no one is in the area.

Most old houses have a keystone or lintel stone over the hearth, and other stones much larger and heavier than other blocks in the chimney. Be especially careful when removing these. Be sure to have as much help on hand as needed to safely complete this work.

Tagging The Logs

If you plan on rebuilding the cabin or any portion of it, you'll need to mark your logs and corners so you can reassemble them in the right order and location. This requires some form of numbering system for tagging the logs.

I tagged and marked each log at the corners, with both a letter and number describing its location in the cabin and wall. Each corner was designated by a letter—A-B-C-D—and the each log with a number.

Start by designating one corner "A", and successive corners B, C and D as you move around the building. Then, beginning with the bottom log in each corner where they intersect, mark the log ends in ascending order on the outside of each notched end, A-1, A-2, A-3, A-4. All the logs on one wall will have even numbers, the other odd numbers.

The combination of letters found at both ends of a log—A-B, B-C, C-D, D-A—will identify which wall it belongs to, and the numbers like—B5 or C8—will tell you where exactly it belongs in that wall.

For the breaks in logs which occur for windows and doors, I used the codes for that wall, the number of the break in the wall, and a letter. For instance, the bottom log in first break for a window in wall C-D was marked C-D I 1, while the log on the opposite side of the window frame was marked C-D II 1.

Use whatever coding system makes sense for your project. Be sure to include how your coding system works in your notes on the cabin.

As far as the actual marking or tagging of the logs, use your imagination and what's available. Do not, paint or carve your markings directly on the logs as it will only create more work for you later.

If you have some way of stamping your codes on tags, you could nail these tags directly to the logs, as I've seen done. I tagged my logs using strips of fiber webbing tacked to the end of each log, and painted with a number. Any durable material you can mark with paint or marker which will not fade for at least a year will do.

After all the logs are marked take photos of each corner, and along both sides of the window and door frames.

If you want to rebuild the floor, loft or roof using the original rafters be sure and mark these as well at both ends, so you can easily slip them back into the appropriate place. Again, take pictures.

Take notes of everything as you go, including an explanation of your marking system, and anything unusual about the construction which had to be tagged.

When you think you are through, go back over the entire cabin and look for anything you missed which should have been marked.

Once the floor deck was built, rebuilding the cabin shell was just a process of matching corners and aligning the walls.

PREPARING TO TAKE DOWN THE CABIN

Logs marked, cabin well documented, there's still a little more planning before the work can actually begin. Ideally, you should allow yourself all the time needed for a careful dismantling of the cabin. That's not always practical, however, especially if an owner is eager to get the building out of the way.

But before you begin dismantling the cabin plan how you will move each of the logs once they are down; how you will transport them to your building site; and where or how you will store them until ready to begin reconstruction.

Completely clearing the work area around all sides of the cabin will make it easier to dismantle the cabin in an orderly fashion. You'll have more room to work, and can situate a truck or trailer as near the cabin as possible for easy loading. Try to avoid dragging the logs any distance, as their surface may be easily marred. If you must move them some distance from the site before you can load them use shortened sections of round log as wheels(See Chapter 2, Part II)

Taking down, moving and loading logs is strenuous work, especially if the cabin was built of hardwoods or large logs. It's easy to injure yourself when loading or unloading logs. Make sure you have enough help to move them carefully. If available, use some combination of a come-along or wench and log wheels to make the job easier.

If you don't plan to use the logs immediately, it's critical that they be protected against prolonged exposure to moisture and insects. A large barn with a free flow of air offers an ideal storage place. If one isn't available consider building some form of shelter, or invest in large tarps to protect them. If you cover them in the open with plastic which can not "breathe" check them regularly for signs of condensation moisture.

However you store them, stack the logs at least 16 inches off the ground to prevent against insect infestation. You can make a frame for stacking them by laying 6x6 beams, or equivalent logs, across concrete blocks spaced every eight feet apart. Stack your logs so the air can circulate freely around them, and place a piece of tin or metal where the bottom log contacts the supports.

If you plan on putting the cabin up immediately, or will store your logs on the building site, make separate piles for each wall. Also, remember to sort then stack the logs in reverse order of how they will be used. The logs for the lowest course should be on top of the pile, and those for the top courses at the bottom of the pile. This way you'll be able to retrieve the logs as you need them without additional work sorting through the pile.

FINAL PREPARATION

Safety should be your first concern as you prepare to do the actual work. Take advantage of any offers to help; there's too much work involved for one man or woman. It is your responsibility to ensure the safety of yourself and anyone else involved in this project.

The size and weight of materials used in a log cabin could easily cause serious injury. Everyone should wear protective head gear, safety goggles and work gloves.

As far as tools go, a heavy duty crowbar or prybar will prove indispensable. Look for one between 24 and 36 inches with a curved end and a flat end. In addition, you want a selection of smaller tools for prying logs or pulling nails. The tipped end of an old tire iron can prove handy, or a large screwdriver. Make sure there's a hammer available for everyone working on the project.

Then proceed with caution. The removal of one log or beam, even a part of a door or window frame, could cause an entire wall to suddenly collapse. Think about the possible repercussion of every action before you take it.

Undoing The Roof

Before you can take down the logs you've got to get the roof out of your way. Because of the dangers involved its a good idea to have at least one helper around to assist you in dismantling the rafters, and move materials safely out of the way. Wear helmets.

When dismantling the roof, start by completely removing the roofing material, leaving the lathe or sheathing, roof rafters, and ridge board exposed. Then, beginning at the uppermost and outermost corner of the roof, remove the lathe. Work down one side of the roof, then the other until the first set of rafters is free standing. If you are concerned about standing on a weakened roof, you should be able knock or pry the lathe loose from below with a hammer and crowbar.

Be careful as you take down the rafters: once the lathe is removed, they can collapse under their own weight, especially if they are not nailed to a ridgeboard.

Once a pair of rafters is free standing, pry them loose from each other where they meet or at the ridgeboard, then loosen them at the wall enough so you can gradually lower them toward the floor. It's easier to swing the rafter slightly out and away from the center of the room as you bring it down. Carefully repeat this process, pair by pair until the entire roof is safely down and out of the way.

When removing a roof in which the rafters are nailed to a ridgeboard, remove all rafters save those in the middle and at both ends first. Then, nail boards supporting the ridgeboard to the wall of the cabin before proceeding. For this you can use one of the rafters removed earlier. Once the ridge board is supported in this way at both ends you can proceed to take down the last of the rafters.

Taking Down The Log Walls

While taking down the logs try to avoid climbing on the walls themselves. Always work from a ladder positioned near an inside corner, or near the middle of the log as needed.

Since the logs used for cabin walls are longer than the inside dimension of the room, you'll need to lower any full length logs down the outside of the building. This is a job for at least two people, one positioned at each end of the log. Whether you raise the logs with some from of machinery, roll them off the side, or raise and lower them with a tripod, the initial steps are the same.

First remove any chinking from the gap between the log and the log beneath it. Again, inspect the log for any potential weakness which might cause it to break as it is being moved.

Also, examine the notches for any pegs or spikes holding the corners together. If there are, remove them. Then, slip the flat end of your crowbar between the notched logs. Apply enough pressure on the log to lift it an inch or so into the air. Work slowly and carefully as the notches can be easily damaged if any of the wood is soft.

When lifting the logs at the notches, use your prybar and some combination of wooden wedges, as you usually won't get enough leverage from the prybar alone to clear the notch. Whenever you rest your prybar against a log for leverage, be sure to insert a piece of one inch board between the log and contact point; without it the force of the bar will "dent" the soft outer layers of the log.

If the log won't move freely, remove whatever is holding it to the log beneath it. If you can't pry or pull the pike or speg freely from above, raise the log enough to slip a saw or hacksaw in between the logs to cut it free. First raise the log, and slip in a wedge to give you space to work a saw blade between the logs.

If you cannot raise the log at all, you will need to devise some way to pull what's holding it from above. With wooden pegs, try drilling a hole into the peg, driving a nail into the hole then pulling it and the peg free. Where spikes are used to hold the logs together, you may need to chisel enough wood away from the head of the nail to slip in a hammer or nail puller.

With stubborn notches, raise log end with a block and tackle or other lifting device and pry it free, gently, from the log beneath it.

Lowering The Logs

If a tractor or some other heavy equipment with a lift is not available, you can adapt the tripod methods discussed earlier for raising your logs and then lowering them safely to the ground.

Another alternative is to set and secure large beams or logs against the outside of the cabin on an incline. Then, slowly roll the logs down to ground level where they can be moved out of the way. For safety, tie pieces of long rope to the wall near both ends of the building and lope them around the log near both notches. This way, you and an assistant can gradually release more rope to control the speed at which the log rolls down the ramp.

Only as a last resort should you raise the logs and let them fall to the ground. This is dangerous and it is likely to damage some.

Windows And Door Frames

If possible, leave the window and door frames intact as long possible, as they provide support for smaller sections of logs. Work down the frame, pulling the nails or pegs holding each section of log. If necessary, cut the logs free by inserting a saw in the narrow gap between the log and frame.

On older cabins, where pegs hold the frames to the cabin you may want to preserve them intact. If so, you will need to remove the frame first to avoid damage to the pegs.

First nail a board along the outside or inside of the wall to support the logs, then dismantle the frame. Start at the top, and slowly knock the frame loose, once section at a time.

Remove the horizontal top section first, then knock the outer sections in toward the center of the frame. Work slowly, with measured alternating strikes along the outside of the frame from top to bottom. When striking the frame with a hammer, place a board between the hammer and frame so you won't leave the mark of the hammer.

REMOVING FLOORS, JOISTS AND SILLS

Many old cabins have several layers of floors, carpet or linoleum which must be removed. If planning a historical restoration, you will want the boards from the lowest, original layer of floorboards. These are usually made of rough planks or even hewn logs with considerable gaps between them. They may be held in place with cut nails or pegs.

If you do want to recreate that authentic cabin you'll want to work slowly when removing the old planks. Also try not to bend or break the nails or pegs holding them in place as you may want to reuse them for that historical accuracy.

For anything other than an authentic restoration, I recommend building an entirely new floor. It's much easier than messing with old flooring, especially if the cabin features any type of tongue and groove floorboards. It's nearly impossible to remove these boards without damaging some.

Until you actually start taking up the floor you won't know what you can salvage. Often one layer conceals considerable damage from moisture or neglect in the past in the layer below. Work carefully with your crowbar to pry the boards loose, one layer at a time.

Unless the floor rafters or beams or logs have been damaged by moisture, you should be able to reuse them. If you want to re-frame the floor the same way, mark the floor rafters and sills with an identifiable code so you will know where they belong.

But before you plan on using them, be sure to measure the gaps between these floor supports. Some cabins were built with gaps as much as three feet between rafters, resulting in a rolling or dipping floor with uneven support. The rafters supporting modern floors are usually on 16 inch centers. You'll want that strength.

PUTTING IT ALL BACK TOGETHER

If you've been careful about dissembling a cabin, you'll find putting it back together a breeze.

In fact, the toughest part of re-assembly may be building the floor to the right dimensions so the walls will fit into place around it.

You have several choices: use your old sills and floor rafters to re-frame the floor; set new sill and end logs in place and notch or nail your floor rafters; or build the floor as a free-standing deck and rebuild the cabin around or above it(See Chapter 2, Part II).

Once you're past this challenge, rebuilding the cabin is a matter of dropping the pieces in place. All your logs and framing timbers are numbered so you know where everything fits. You have detailed notes and photographs for reference when any questions arise.

You should take the time now, before the cabin walls start to rise, to plan and prepare for the running of modern amenities like electricity, plumbing and heat ducts, if required.(Again, refer to Chapter 2, Part II and Appendix B: Adapting The New). You'll find it much easier to prepare for these now than to try and work whatever is needed in after the log walls are up.

After you've addressed those concerns, put the cabin back together in the reverse order of how you took it down. For safety and to protect the logs, you'll need tripods or some other solution for slowly raising the heavier logs and setting them in place.

Start matching your codes in one corner and work your way around the building, one course at a time. Everything should slip into place, but you may need to roll or twist some of the logs slightly for a tight fit.

If you're not striving for a historical reconstruction you'll be better served using nails of varied sizes where wooden pegs were used in the past. Also, you'll find it easier to use standard lumber when framing windows and doorways.

If the original cabin featured a loft you may need to improvise when setting the beams or logs supporting it in place. Original log cabin lofts often featured some measure of roll to them. That may not be acceptable in your plans for the space. If the notches in the logs which hold the rafters don't line up precisely you won't have a level loft. This can be easily corrected using a level, saw and chisel to rework the notches to the appropriate depth and alignment.

Once the walls are up—and it can take as little as a day or two—prepare the uppermost logs to serve as or hold the plates for the roof rafters.

REPLACING DAMAGED LOGS

Frequently, the condition of one of more logs from the original cabin forbids their use in the new structure.

Replacing an entire log will not present any problems as long as you use logs of the same width, height and length. The vertical height or the face of the log is most important. If you use a log too small or too large, and don't compensate when cutting the notch, it will throw the rest of the cabin off.

For my replacement logs, I used logs from walls in the loft of the original cabin since I wasn't planning to recreate it. If there are no "spares" available to you, you'll have to hew out a log of the required size. Avoid using a fresh cut green log if possible, as it will shrink over time and could introduce new problems.

When notching a replacement log, you want to match the size and depth of the original log, exactly. If possible, use the notches from the log you are replacing as a pattern for the cuts.

If that original is not available, you will need to match the notch cuts to those in the logs below and above it. Work on the bottom cuts first, at both ends, then match the top cuts. This may mean raising and lowering the logs, but it's absolutely necessary that you get the right tight fit for the reconstruction of the rest of the cabin

Marcus secures the block and tackle as we prepare to raise a log, then helps line it up with the wall before we set it in place.

Adding A Loft for Upper Rooms

In cabins in which the loft was an original feature, there's usually a pair of larger, stronger logs on opposite sides of the room which support the loft's floor rafters or beams. Use these same logs to support your loft, even if it means rearranging the order of logs, and trimming notches so they run in the course at the desired height.

More often, though, the loft was an afterthought, added years later to an original cabin to add space for a bedroom or storage. The rafters were notched into whatever logs were at the desired height. These may not be strong enough to support the weight of a modern floor. In that case you'll need to replace them with larger logs from other parts of the building, or hew out entirely new logs to support the floor.

When building a loft, you can relieve pressure on the log walls, and strengthen the loft, with an interior wall, or by attaching posts to the rafters supporting the loft floor.

Windows, Doors, Interior Walls

Even when trying to recreate an authentic cabin, I recommend using modern methods and materials for windows, doors and the roof.

The biggest challenge, however, may prove locating the doors and windows to fit the old spaces. There's nothing standard about a real log cabin. If want to keep to the original dimensions, you may need to build doors and windows to fit.

As an alternative, enlarge the size of the openings to accommodate standard doors and windows, or fill in the frame with additional lumber on one or more sides, to accommodate a standard door or window. Whichever approach you take, you'd be better served by deciding on your solution before the walls are up and built.

FRAMING THE ROOF

Once the top logs which will support the roof rafters are in place, I strongly recommend you employ modern framing methods and tin roofing as outlined in Chapter 2 of Part II. Before you can begin, you may need to build up or bring down the surface of the top logs which will support the rafters so they are even with each other.

For an authentic reconstruction you can use the original rafters, or replace them with the saplings, or hewn beams of comparable size. For an authentic period roof, you'll need a froe and ample supply of oak or cedar to split new wooden "shakes" for roof shingles.

I've never attempted this—nice as they look they can be a fire hazard. If you want a shake roof I advise you consult the original *Foxfire* book for detailed instructions on building a shake roof. It's the best source I've seen on how to do this.

MODIFYING OR ADAPTING AN OLD CABIN FOR NEW USES

Most of the log cabins I've seen taken down and put up again are never entirely true to the original. Some of the original form is still there, but it's been somehow modified or adapted to the needs of the builder.

In my case, for example, I took down an old cabin, but only three walls of it were salvageable. So, I added three walls onto an existing cabin and ended up with a new room that looks just like the original cabin, on three sides.

I've also seen the notched ends of logs cut off to build smaller cabins than the original. And just the opposite: The logs of several cabins put together to create an entirely new structure more ambitious and spacious than any of the cabins which were the source of its logs.

Once a cabin is dissembled it's logs can be adapted to any purpose. You'll lose something in length to the process, but by combining several cabins you have many more options in what you build.

Here's a couple of ideas for how logs can be combined for a greater running length.

First, you need logs which are a close match in width across the vertical face and thickness. Using logs of the same type of wood is preferable but not absolutely necessary.

The simplest method is merely to lay them squared end to squared end with their butt ends touching, toe-nailed into each other. They can be propped up with a support resting on the log beneath, and by nailing a short length of board or trim on the outside of the logs where they meet.

A sturdier method is to join them where they meet with a modified version of the lap notch. In this method, an area eight to 12 inches long, and half the height of the face of the log, is removed from the top of one log and bottom of the other log where they meet. Slip the ends of the logs together, secure them in place with spikes or nails, and add some support beneath this joint.

When you want to build a new cabin from old logs, its usually best to cut all your notches fresh. Trying to match up notches which weren't paired to begin with can introduce a lot more work into the process, even if they are the same type of notch.

For interior walls in a large log home, you can use 2x4s and drywall to frame the walls and provide additional support for the ceiling and upper rooms. If you have enough logs, you may use them to construct the dividing walls. You'll save a lot of work and headaches if you notch these logs into each course of the outer wall as they go up. A flat saddle notch, slipped into the gap between logs, offers the easiest solution.

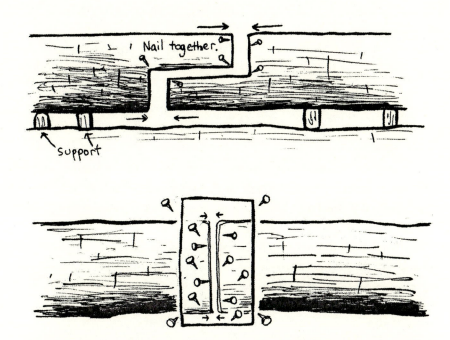

Two ways to combine shorter pieces to make one long log. Top, cut a variation of lap joint so pieces fit together, then nail and support as needed. Bottow, butt log ends against each other in wall, toenail together and nail a board as trim to hide and strengthen joint.

CLEANING THE LOGS

Old log cabins are usually covered with newspaper, wallpaper or whitewash, and the dust and grime of decades. Cleaning the logs, and then treating them with some form of preservative are important steps in the reconstruction process.

I've found it best to tackle an entire wall at a time, once it's up, than to clean each log separately on the ground. If the logs are still sound you can use a hose and water to remove much of the dirt, grime and shards of paper.

You'll need to employ some elbow grease, nylon brushes, a mild soap like Murphy's Oil, and plenty of water to get them completely clean. Do not use a steel brush as it will leave marks all over the face of the log. After you've scrubbed them clean, rinse or hose them off and give them ample time to dry before applying any type of oil, finish or preservative.

I never had to deal with logs which were painted with anything but whitewash, which is easily removed. I have heard of others who had to strip layers of paint from their cabins before they could see the original color of their logs.

For this, a sandblaster was recommended. Instead of the usual grit, though, finely ground walnut hulls or corn cob hulls can be used during the sandblasting to remove paint without damaging the logs. (Available from the Perma-Chink company; check their website or catalog.)

Treating The Logs

There's a couple of ways to treat the logs once they are up and cleaned. I've always used a mixture of linseed oil and a water sealant to bring out the grain and provide a protective coating against the elements.

Occasionally, some old logs may hold areas with of damaged soft wood from simple decay or as a result of dry rot. You must remove the infected wood or it could damage the surrounding area of the log. Chisel out the soft wood until you get down to the hard wood. Treat this section with a wood hardener, following the manufacturer's instructions.

For soft areas of wood, I've used the Min-Wax wood hardener. You can brush this on or inject it into the wood and it hardens to something approaching the original strength of the log. Other products of this type are also available.

After the wood has dried and hardened, fill in the remainder of the section with a wood putty or epoxy to restore the shape of the log. When it dries, you can paint it to match the rest of the log.

For the interior of the last cabin I restored I used a polyurethane stain which brought out the grain of the wood and dried into a protective coating. By the time you read this there may be other options in wood protectors and stains so check with your local hardware supplier and other resources.

When applying any stain or finish, wear the appropriate protective gear and make sure the cabin is well ventilated. Give the logs ample time to dry—at least a few days—before you use the space.

CHINKING

Chinking the old log cabin is no different than with any other cabin. Follow or adapt the methods outlined in Chapter 2 of Part I. The one advantage to chinking older logs is that they will not shrink over time. You'll still need to check the chinking from time to time, and touch up as needed, but there's not a lot to do as the logs have already finished shrinking.

FINISHING TOUCHES, FINAL THOUGHTS

Log cabins, as they were originally built, tended to be dark affairs, with a minimum number of windows. They were also often cramped quarters with low ceilings. This environment will work fine for those who want to accurately recreate this humble home of the past.

Today, though, we have different means, and different needs, and it's important that you address them. Adapt to the needs of all who will use this space. A log cabin can still be authentic with electricity and plumbing, its ceiling raised and windows enlarged or added to let in more light.

The log cabin has always been a symbol of individualism. It should be a statement about you, what you value, the qualities of life you cherish, whether you build it your self, or build on the efforts of someone who shared the same ideals in another era.

It's the finishing touches, the personal flourishes,—what you bring and give of yourself to the structure—which make any house a home.

Epilogue

If the preceding pages convince you of nothing more, they should establish that I cherish the log cabin and all it represents. There's just something about its rustic simplicity which speaks to my heart and my admiration for the individualism I consider one of the American virtues.

But I'm a realist too, and, recognize there are those who do not hold the log cabin in such esteem. In fact, to many of the old timers here in Tennessee, the log cabin is little more than a symbol of the hardscrabble poverty they worked so hard to escape and forget. For them, progress and success is a nice brick home.

"Ain't you gonna cover them up," asked one astonished local woman when we showed her the work we had done uncovering and restoring our one room cabin.

"Whah don't ya paint them logs green?," suggested another who had also spent her earliest years in a log cabin.

No, not everyone shares my romantic notions about log cabins, and at times my wife Liz is among them. Although we have worked side by side for more than 25 years, and literally built our home ourselves, she has been quick to point out the shortcomings of log cabins. Often her comments are prompted by some turn in the wind or weather which highlight the same reasons previous generations were often so eager to leave a log cabin in their past.

In fairness to her, and them, and to any of you who are considering building a cabin, I'll address here some of the most common complaints.

Log cabins are dark: There were precious few windows in the original cabins, which meant they were exceedingly dark most of the time.

There's an easy remedy for this, one I used in the cabins I built myself: Add as many windows as possible, even a skylight.

Log cabins are drafty: As the logs check and shrink, gaps appear in the logs and chinking which give the wind easy entry into the room.

You have to keep after them, and regularly repair or touch up the chinking.

Log cabins are cold: In winter weather, the cold wind which slips into the room around these cracks can work against any effort to maintain a consistent level of heat.

Log cabins require continual upkeep: I'll venture to say that this complaint applies to any home, and especially one built of wood. With a log cabin, though, you must constantly tend to the chinking, plugging the gaps and adding new mud as soon as they develop.

If you do, you can eliminate the drafts, and keep out the cold.

Log cabins can be dusty and dirty: And what house won't be, if it's neglected?

In fairness though the logs seem to attract dust which must be wiped away regularly. Over time the chinking, especially, generates its own dust as it ages and crumbles.

The Work is Never Done: In this, a log cabin is really no different from any other do-it-yourself home.

There's a tendency, though, once the walls are up and chinked, and the roof is over head, to move into the space before it's fully finished. Once inside, its easy to put off the final touches.

Add to that the regular the miscellaneous upkeep every home demands and your log cabin becomes a long term project, at least in small doses.

In closing, I offer that these are all really minor issues, inconsequentials easily remedied, which should be recognized and accepted as such.

The rain will fall, the wind may blow, and the cold can force its way around the door.

Whether it last a lifetime or a century, the log cabin you build yourself will stand as a testament to the patient determination, personal pride and ideals which inspire you to shape your home from the woods.

All things considered, it's a nice place to live.

Appendix A
Tools

You could build a log cabin, like many a pioneer who braved the wilderness, with little more than a good axe.

There's no reason to, however, unless you crave that authentic experience. In fact, the more tools you use, the easier the job becomes. But, you may be surprised at the range of tools required for the job.

Before you start work on your cabin project, take time to familiarize yourself with the tools which will make your job easier. What follows are brief descriptions of the many tools you may employ when completing a log cabin project. They are grouped by purpose, and introduced in the order is which you can expect to need them. Some you'll reach for again and again throughout the project.

Except for a few of the specialized hand tools for working with logs, all are readily available at local hardware and home improvement centers. If you can't find the specialized tools at yard sales or flea markets, try eBay on the Internet, or the Cumberland General Store in Crossville, Tennessee or the Lehman's mail order company listed in the "Resources" section.

I've tried to make this a comprehensive list. Forgive me if I've overlooked anything.

SAFETY FIRST!

I can't over-emphasize the importance of protecting your safety when building a cabin. You'll be chopping, pulling, raising, lowering, and cutting logs with many tools which can cause serious injuries.

Only you can ensure your safety and you *must* plan for it. Err on the side of caution. Always allow yourself ample room to work. Be aware of others, and make sure they are aware of you and what you are doing.

Get and use reliable, durable safety equipment.

Here's the basics:

Gloves: Expect to wear out your work gloves over the course of the project. Make sure they offer a comfortable but snug fit, with leather to protect your palms and fingers. Personally, I prefer the combination leather/canvas gloves with a heavy cuff over the wrist.

You must wear them if your gloves are going to do any good. I almost lost a finger once when I slipped while carrying a sheet of tin, gloves folded in my back pocket. My finger had to be stitched back together and I couldn't work until it healed.

Wear your gloves.

Safety Glasses: An absolute necessity when cutting, chopping, hammering and chiseling wood. I prefer the wraparound style which fits like sunglasses. Whatever your preference, get a pair that fits tight, and won't steam up and blur your vision when it's hot or when you move from indoors to outside in the cold.

Ear Plugs: Unless you want to spend your golden years in silence, wear ear plugs whenever you're working with power tools. Options include "buds" which slip into your ear, wax plugs, and padded headphones. You can also get earplugs attached to a cord you wear around your shoulder. I always have a tissue or ball of cotton in my pocket to stuff in my ears if I lose my plugs.

Shin Guards: Advisable when you're squaring off logs with an adze, as the chips will be flying as you swing the blade toward you. I recognized their necessity when my adze glanced off a log and into my leg. Don't repeat my painful mistake.

Soccer or baseball shin guards should offer enough protection and aren't that expensive. I made my own "chaps" by slipping Styrofoam inside leather pouches which I tied around each leg from the ankle to knees.

Your shin guards should provide a snug fit, yet be loose enough to allow you to move freely.

Helmet: A helmet is another absolute necessity whenever you are cutting trees, raising logs or working anywhere something could knock you in the noggin. It's simply not worth the risk of not wearing one. Any protection is better than none at all, even a bike helmet will help.

A solid lightweight, plastic construction hard hat may be your best bet. Make sure it's comfortable so you're never tempted to set it aside, even for a minute.

Dust Mask: Uncovering the logs will raise dust from the old chinking, and any nests you uncover can harbor disease. This dust can be irritating at best, or a potential health hazard, at worst. An inexpensive dust mask, worn over your mouth and nose, provides a simple solution.

CUTTING THE LOGS

There's only a few tools required to bring down the trees, cut your logs and move them from the woods.

Chain Saw: A chain saw is a real work saver. You'll be doing a lot of cutting with the saw so get one you can handle comfortably. Look for safety features like a safety no-kick tip and automatic bar brake. Learn how to sharpen the chain properly.

A chain saw can be a very dangerous tool in the wrong hands. If not confident in your ability, consider having some one who is experienced cutting down the trees which will make your logs, or take the slower approach using a cross cut saw or axe.

Cross-Cut Saw: Before there were chain saws there were cross cut saws. They feature long blades and large teeth for cutting trees into logs. Cutting with a cross cut saw is a good workout.

A two-man saw, with handles at each end, can slice through a log surprisingly fast, once each masters the rhythm. There are also one-man versions around, with two handles at one end.

You'll need a saw "set" and file in order to keep it sharp.

Bow Saw: A hand-held bow saw can prove helpful removing branches from your logs and for "scoring" notches or small sections of log. A bow saw with a 21-inch blade will prove adequate, though they come up to 36-inches long. Make sure the blade stretches tight with no wiggle room, or the saw will bind in the cut. Keep a spare blade on hand.

Axe: You can bring down any tree with a long handled axe. It's just a lot of work. Choices include single bladed or the more versatile double bladed axe. An axe will also speed up the process of trimming unwanted branches from trees and some of the finishing work on your logs.

Learn how to handle the axe safely, sharpen it properly. As with all hand tools, always have a spare handle on hand.

Wedges: When a tree hangs, binds your saw or starts leaning the wrong way, wedges can help set things right. Get several, of different sizes. Iron wedges are the strongest but wooden or plastic wedges eliminate the risk of damaging your cutting tools.

Make sure you have a large hammer or hand sledge on hand to drive the wedges in place, as needed.

Cant Hook or "Timber Jack": Once the tree is down and you've topped it, you can use this handy tool to raise the log up or roll it around. Later it will really prove indispensable when you're ready to start working on the logs.

Prybar or Tamp Bar: An iron rod about six feet long, sometimes with a chisel blade at one end. A useful aid for lifting, moving and rolling logs. One of those tools you'll find plenty of unexpected uses for as you work on any cabin project.

Come-Along, Block and Tackle, Wench: All pulling tools which, when used in combination with steel cable or chains, can help guide a tree to fall where you want it to, or for moving logs through the woods. A block and tackle or wench is also useful when raising the upper logs and beams in your cabin.

Chains and Hooks: Heavy duty chains, with hooks on the end which you'll attach to logs for dragging through the woods. You can expect to need more than one chain, and its advisable to have some variety of chains of different lengths as well as tow hooks, for moving the logs.

Tongs: Attached to a chain, these grab and hold a log for dragging to where you need it.

WORKING THE LOGS

When it's time to work your logs—hewing the sides or cutting the notch—there's a number of more specialized tools you'll need, as well as your chainsaw or bow saw, chains, cant hook, and prybar.

Log Stands: Once you've dragged the log to where you want it you'll need to improvise some form of stand to hold the log in place so you can work on it safely. You want stability and plenty of clearance so your axe or saw won't strike the ground.

For instance, I used a wide sections of log, flat on top and bottom, at least 12 inches in diameter and 18 inches high. Just lift both ends of the log and prop it on these supports as needed, then drive in wooden wedges to prevent the log from rolling.

A modified saw horse or stand can also be used.

Bark Peeler: Although there's a tool called a "bark spud" for removing the bark, any wide flat-blade chisel will do. With some types of trees, like poplar, once they are well seasoned removing the bark is as easy as peeling it away.

Axe: There are several types of axes you may use when working with your logs.

Any single or double bladed axe with a long handle and wide, sharp blade can be used for "hewing"—cutting away the outer surface of the logs to square them off.

A broadaxe is specifically designed for this work and will really speed the job along, leaving a nice, clean face to the logs. They come in several and styles, with blades 10 inches wide, even wider.

A broadaxe is much heavier than a regular axe, and harder on the back and arms. The handle should curve away from the face of the log to protect against skinning your hands as you work along the log.

For notching, I recommend a broad hatchet with a blunt end which can be struck with a hammer, or a wide bladed chisel, at least four inches across.

Also, add a bastard file for sharpening your axe to your tool box. Learn how to properly sharpen your axes.

Adze: Once you master working with an adze your logs will look as smooth as if they were squared at the saw mill. That can take years, however.

An adze features a curved blade which you work toward you along the horizontal face of the log. This can be both backbreaking and dangerous work as the chips fly up with each swing of the blade toward you.

Like an axe, the adze comes in several sizes. You'll want to find something you can comfortably hold and swing. The handle should be long enough that you don't have to bend forward to work. Be sure to keep a spare handle on hand.

Chisels: A selection of chisels, from one-to four-inches wide will prove handy for removing stubborn bark, smoothing knots and cleaning up your notches. Get ones which are all metal, or have an acrylic handle for striking. Wooden handles tend to break too easily.

Raising and Setting The Logs in Place

Consider yourself fortunate if you have a tractor, fork lift or other form of front end loader with which you can raise your logs and set them in place.

If not, you will need to adapt available tools and methods to the challenges of getting the logs off the ground and onto the structure.

Tripods: Without heavy equipment or a ramp and the horsepower to roll your logs up and into place, you'll need to construct some form of tower for attaching your hoist and raising the logs.

A simple tripod does nicely. You can build one by lashing sturdy saplings together. If you don't know how to lash, consult your local Boy Scout or the Scout handbook.

The tripod legs should be securely tied together, and there should be ample clearance at top to allow room for you to hoist the logs high enough that you can easily and safely set them in place.

Hoists: You'll want two of whatever type of hoisting device you'll be using, one for each end of the log. Options include an electric wench, chain hoist, block and tackle or come-along. All greatly increase the amount of load you can raise comfortably and safely. With an electric hoist all you need do is attach your logs and push a button.

Whatever you choose, make sure it features an anti-slip lock so there's no chance the log can suddenly come crashing down on you if the equipment fails.

Small Sledge: If you plan on spiking logs together, a 3 to 5 pound short handled sledge or heavy duty hammer will give you the power to drive those spikes into place.

Crowbar: Crowbars, prybars or nail pullers as they are called in some places, come in a variety of lengths, from barely a foot to five or more feet long. If I only could have one, I'd go with one about 30 inches long. You'll find it gives you enough reach and leverage for most jobs.

Look for bars with a hooked and flat end, with slots at both ends for pulling nails. They will prove handier than you can imagine.

GENERAL TOOLS

Here's a rundown of some of the basic building tools you can expect to use over the course of a project.

Tool Belt: The best way to stay organized and have access to most of what you need, when you need it. A basic tool belt features slots for your hammer, tape measure, pencil, pliers or screwdrivers, and a pouch for nails.

Hammer: An all purpose hammer, as heavy as you can comfortably swing, is requisite throughout the job. Get one with a brightly colored fiberglass handle and you won't waste a lot of time searching when you've misplaced it.

Square: The secret of straight cuts, a square can be incredibly versatile. When it comes time for building the roof, those charts stamped along the face of a roofing or framing square can save a tremendous amount of guesswork, if you know how to interpret them.

Plane: For smoothing wood surfaces and reducing the edge of doors for that tight fit.

Tape Measure: Something you'll use every step of the way. Get a retractable tape measure at least 25 feet long and one inch wide.

Level: A 2-foot carpenter's level should prove adequate for the entire job. You should also have a "line level" on hand, with plenty of line.

Chalk Line: Needed when marking your logs for hewing, or making any extended straight cuts. Stock up on that that bright purple chalk dust, as well.

Compass: A simple compass with pencil will allow you to mark circles as needed, and to scribe logs for notching, if you're using the round saddle notch.

Carpenter's Pencil: A standard carpenter's pencil, well sharpened, makes a nice broad mark that's easy to follow whenever you need to make a cut. They're cheap and easily misplaced so buy several.

Hand Saw: On every project you'll always find some situation where the simple hand saw is exactly the tool you need. Keep it clean and free of rust.

Screw Drivers: Flat headed screwdrivers of varied lengths will prove handy when you've got a stubborn board to remove. I keep a bucket with several different lengths handy, as well as the lug wrench/hub cap puller which used to be standard with car jacks. Improvise if you need to, with whatever tools you can use to loosen, raise or remove what's covering the logs.

Folding and Extension Ladders: You can build your ladders, buy one of each, or invest in one of those multi-purpose convertible ladders.

You'll need an 8-foot folding ladder and a 16-foot extension ladder before you're through.

For Roofing: Tin snips; roofing square; metal blade for circular saw; roofing hammer if setting shingles; hex nut head if screwing down tin.

For Electrical: Assorted wire cutters, strippers, crimpers and pliers; circuit tester.

For Plumbing: PVC pipe cutter; assorted wrenches, pliers and screw drivers.

POWER TOOLS

In addition to the chain saw, several other power tools can make your life easier over the course of the project.

Drill: You will need to drill some holes, for bolting door and window frames to the logs, or borings hole to run electrical lines. You can use an old fashioned man-powered hand drill or brace, but a power drill is so much faster. Rechargeable battery operated drills can be a real convenience.

In any case, you'll need a selection of interchangeable tips in several sizes up to 5/8-inches, at least.

Circular Saw: A 7–1/4-inch electric circular saw will really move things along whenever you need to cut your framing lumber or flooring. Always have several spare blades handy.

Chop Saw: An electric miter saw or chop saw will save a tremendous amount of work and make sure you've got the perfectly straight edge for every cut.

Jig Saw: For fancy cuts where your circular saw won't work.

Reciprocating Saw: The "saws-all" is like a larger version of a jigsaw, but much more versatile. Useful for everything from fixing notches to cutting space in the wall for another window. Get plenty of blades, of varied sizes.

Electrical Cords: Heavy duty extension cords, the more the better.

CHINKING

I am convinced plain old mud, as originally used, makes the best chinking because it's so easily repaired, and chinking always requires patching up.

For more permanence you want to use some variation of concrete mortar. Regular mortar works fine for narrow chinks but may not hold up on some of the wider gaps. I used regular Quickcrete mortar cement for my first cabin and its still holding up after nearly 25 years.

With the amount of chinking required for an entire cabin, you'll save a lot of money mixing your own "mud." I asked around about what's best for chinking as was advised to use a 1 part Portland cement, one part mortar and five parts sand. It's worked fine for me. You'll need to improvise a little for the right mix.

Others have recommended a product called Perma-Chink from Perma-Chink Systems to me because it expands and contracts with the weather, eliminating problems associated with shrinkage. It was always beyond my budget so I've never had the opportunity to use it, but you may want to contact the company or visit their website

Incidentally, if you want, you can buy powder you can add to your mud mix for the look of real mud. Ask at your concrete supplier.

Cement Mixer: You're going to need more mud than you can imagine to chink your cabin. The mud mixer can be you, or a machine. You <u>can</u> stir the ingredients for your chinking in a kiddie pool as I have. It's backbreaking work which really slows you down.

I finally invested in a "portable" electric cement mixer and saved countless hours on my varied log projects. You still have to shovel the ingredients into the mixer, but it's not near the chore of mixing mud with a hoe.

Mixing Pan: If you can't get an electric cement mixer, you'll have to mix your mud by hand. A plastic kiddie pool works fine. Get a pan or tray large enough to allow you to mix enough mud to fill between two and three wheel barrows. Or, you can use the wheelbarrow itself for mixing one small batch at a time.

Mud Board: A piece of board used to hold the mud as you slide it between the logs. I used a piece of 3/8 inch plywood about 18x24 inches. The size and shape depends on what's comfortable for you, and how much mud you want to haul and handle between refilling the board.

Trowels: The broader your selection, the easier it will be to push your mud in place. I recommend one large and one small triangular mason's trowel as well as a square hand trowel. I've used all three chinking the same log.

Whisk or Hand Broom: Unless you want a shiny finish to the mortar, you'll need to brush its surface after it starts setting. A whisk broom works fine but any stiff brush will serve.

Caulk: When you're working with relatively new logs, they will shrink away from the chinking during the first few years. I've used clear acrylic caulk to seal the gaps which develop. Once it's dried you can't really tell it's there. Perma-Chink also offers several products which match the look of real mud chinking.

Wheelbarrow: A good sturdy model will enable you to move your mud to where you need it.

Wire Brush: A stiff wire brush is needed to brush away excess chinking and remove it from logs after it's had a chance to dry a while. Be gentle, or you'll scar the logs.

Cleaning Logs in An Old Cabin

Buckets: Several 5 gallon buckets to hold the water for scrubbing, cleaning and wiping the logs.

Brushes: An assortment of scrub brushes with and without handles to get the dirt off. For most work a nylon or bristle brush will be adequate. A wire brush makes it easier to clean up stubborn areas where the chinking or paint clings to the logs, or the joints around the notches. Be careful not to work the wood too hard with a wire brush, though, or you'll leave scars.

Chisels: For removing bark or damaged wood as needed. Remove the bark on logs before chinking with a wide blade chisel; a standard chisel with a 1–1/2 inch blade should serve wherever you have to remove small pockets of rotted wood.

Broom and Shovel: Cleaning the logs creates a real mess. A good stiff broom and shovel will make it easy to keep your work area clean.

Cleansing Agents: Straight water does a surprisingly good job. If the logs have been painted or are particularly filthy try scrubbing them down with TSP and water before resorting to anything harsher.

Avoid toxic solvents as much as possible. If the logs have been painted and you want to remove it, rent a sandblaster, but use something other than sand. Options include ground corn cobs or walnut hulls.

Wet/Dry Vacuum: A wet/dry vac with a large container can be a real work saver when cleaning between the logs and around the site.

PRESERVING AND PROTECTING THE LOGS

Preservatives: Today's log cabin owner has a number of options to protect and preserve the cabin against mildew, decay and insect infestation. Always read the safety precautions on the label before buying.

At the very least, I'd recommend brushing linseed oil on the logs before chinking. Be aware though it will darken the color of the logs. Personally I've had success using a 50/50 mix of linseed oil and Thompson's water seal. You get the benefits of both products in a liquid that's easily applied.

For the interior I've used both linseed oil and a clear polyurethane. Both have a distinct smell, but the smell of the linseed oil will linger much longer after drying. You can get the polyurethane in a range of colors and finishes. I'd recommend the clear satin finish,

Whatever treatment or preservative you are thinking of using, apply it to a test area first. I've always used lower logs near the corners for that trial run. Make sure you can live with the finished look before you apply it to the entire cabin.

APPENDIX B

Adapting The New: Modern Methods and Materials In A Log Cabin

The enduring charm of the log cabin owes much to the rustic simplicity of the structure. Today's builder can achieve that, while taking advantage of modern building methods, materials and amenities.

In fact, adapting these to the traditional log cabin can make your work considerably easier, and the quarters you create more comfortable.

Mix and match, incorporate or adapt what's right for you and you'll end up with a cabin that's more enjoyable and more a reflection of your personal style.

Some considerations:

ELECTRICITY, AND ALL IT ENTAILS

Electricity, from any source, will make the cabin experience more enjoyable. A reliable source of available power opens up more options in how to put the cabin together, and what you can do in that space after it's done.

Unless you're a purist, there's nothing wrong with an electric light, and a few well electrical outlets.

The desire for electricity should shape your decision in the earliest stages of the project, from where you situate the cabin to how you construct its walls.

These considerations are even more critical if you plan to rely on some alternative energy source. Water or wind-driven generators, solar power, even geothermal heating/cooling are all achievable, and more easily so if they shape your plans from the earliest stages. It's always easier to build in your energy solution than build on later.

Running Water, Indoor Plumbing

The only people I've ever heard sing the praises of the outdoor privy are those who haven't brought their plumbing indoors yet. I know, I was one of them.

A reliable water source should be a guiding consideration when planning to build any home, be it a babbling brook or a 500 gallon water storage tank. It's nice to dip a cup from a freshwater spring, it's also nice and convenient to turn on the faucet in winter to draw the water for the morning pot of coffee.

Of course, you can put a faucet outside the door, and build your shower so you can bathe under the stars. We've done all this too. But we like our indoor plumbing. and you probably will too, in some form, even if it's only a single spigot trickling into a corner sink.

Indoor facilities, even that sink will claim some of your space. So you have to plan for it and possibly around it. If winters are hard where you situate your cabin, protect your plumbing against freezing in winter. There's nothing quite as discouraging as lying on the frozen ground when it's 10 below trying to thaw the pipes with a hair dryer. Worse yet are the challenges of repairing the burst pipe, broken and frozen beneath the kitchen.

Think realistically ahead of how you'll use water, where you'll need it and how you can run the plumbing so its easily accessible whenever repairs are necessary. Place cutoff valves wherever practical so you need not be completely without water when making those repairs.

If you want an indoor bathroom, before you get too far in your cabin plans, have a "perc test" done on your land to make sure you can install a septic system and the required field lines.

HEATING AND COOLING

Most cabins had nothing but a fireplace for heat. A month of sitting by ours, our faces scorched while our backs froze, was enough to convince us there had to be a better way.

There is…in fact there are several heat sources to consider. Most one room cabins aren't large enough to warrant a central heat and air conditioning system. If your plans entail a multi room structure, by all means investigate the costs and logistics before you start building. A thermostat controlled heat pump, gas furnace or AC system can make life very comfortable. It's easier, and cost effective, to install any required duct work as the building goes up. If there's decades' worth of firewood on your land, you may also want to investigate a wood furnace.

More likely you'll be adequately served by some form of space heater: a wood stove, gas heater or electric space heaters. Some may also consider a kerosene space heater, but I never liked the smell.

Wood, or gas, you'll need a way to vent the smoke or fumes to the outside. This can be accomplished by building a flue or chimney, and venting your heater to them with pipes. You can also run your pipes through the log walls or ceiling, provided you use insulated pipe. Usually this consists of three layers of pipe, one inside the other, with a gap or insulation between them. In theory the heat never reaches the outer layer. Make sure anything which comes in contact with the pipe is adequately protected, with fireproof insulation if necessary.

Today's wood stoves are quite efficient when compared with those available 20 years ago. You can also buy models with ceramic doors which let you see the fire. The biggest problem with heating with wood is that it's tough to accurately regulate the temperature from room to room. At times it's going to be too hot, at others, usually in the early morning, it's too damn cold.

Gas heat, either fueled by a natural gas line or liquid propane tank, comes close to offering the best of both worlds. The biggest plus is that you can precisely control the heat with a thermostat control so you're always comfortable. The burners in many gas heaters are also exposed behind ceramic glass, so you can enjoy the warm glow of the fire.

Plug-in electric heaters are fine for warming smaller spaces or taking the chill off a cold day. There's many types to choose from, including radiant heaters, ceramic heaters and oil filled electric radiators. Two or more, or one large heater, can be combined to heat a large room.

START RIGHT

A lot of reckoning went into the original cabins, and many a pioneer reckoned the building he put up was near level. Pretty close was good enough.

Today you want to start with a foundation, or series of piers which are perfectly level, corner to corner, side to side. Getting it right can be a slow process but well worth the time and effort involved. If you don't, you'll have uneven floors at least, and doors and windows which won't open or shut, at worst.

Decide first if you want to build the cabin on a continuous foundation, or a series of piers. Then its just a matter of staking out your building, running your line around the perimeter and diagonally corner to corner, attaching the line level, and making those minor adjustments to get it right. Actually, it's more difficult than it sounds and if you're not familiar with how it's done, get a good reference on basic building methods and start reading.

Plan In Multiples of 4

When the original log cabins went up there were no building standards; Pioneer Joe was pretty much on his own.

Somewhere down the line, though, standards took hold which make it much easier to build any structure *and* minimize waste of material.

If you plan your cabin and the rooms in it aware of these standards you'll have a much easier time of it.

The standard measurement for sheets of anything used in a modern home is 4x8 feet. That applies to plywood as much as paneling and drywall. Two sheets laid horizontally give you a wall eight feet long, rising to an eight foot ceiling. For a 12x16 room you'll need 12 sheets of plywood sub-flooring. It works like that all around, if you plan carefully.

Walls, floors and ceilings are framed with these standards in mind. Rafters and studs are positioned on 16-inch or 24 inch centers. That way, every four feet you have something to nail the edge of your first sheet of whatever, and the beginning of the next. Your tape measure probably has special arrows or markings every 16 inches so you know where to put the stud.

Understand how these standards work together and you'll have an easier time planning the structure, and estimating the materials you'll need.

Doors and Windows

Windows and doors come in standard sizes too. There's quite a range, so you should find something to fit your needs. When you build, you want to know the outside dimensions of what you plan to install so you can allow enough "rough-in" space to slip it into place.

Unless you want to build your doors yourself, buy them all pre-hung in a frame you can slip and nail into place with little hassle. I strongly recommend that you install at least one 36-inch door as one entrance to your cabin. It's wide enough to accommodate just about anything you'll ever want to move in or out of the cabin.

Interior doors, room to room, can be small, sometimes as narrow as 28 inches for a room entranceway, 24 inches for a closet.

Closets and Pantries

If you want authenticity, your cabin wont feature any closets. In just about every original cabin I've been in the only "closet" was an oversized nail tacked randomly into one of the logs.

Closets and pantries are nice, and there's never enough of them, no matter how many you build(according to my wife). These will encroach on available space, so plan ahead carefully. If you don't want to permanently box in the space, you can use portable cabinets, wardrobes or pantries. You can easily re-arrange them whenever you want to give the cabin a new look.

Modern Floor Coverings

An authentic floor in a log cabin means rough sawn planks with gaps wide enough to watch the dogs and chickens under the house.

Go modern. Every layer of flooring you put down helps keep more of the cold from seeping in in winter. Be sure to put down some form of sub floor, first.

If it's wood you want, install a hardwood tongue and groove floor over the subfloor. Enjoy the work, though, especially if you chose to build it from 2–1/4 inch wide flooring.

There are other options which work quite well. Linoleum or carpet installs easily and can be bought by the roll or as tiles. Both are easy to clean up . What's more, they are available in a variety textures and styles. A good choice may be the textured linoleum which looks and feels like a hardwood floor.

Add a few scatter rugs, and the place has a homey feel. In fact rugs of any type, even wall to wall carpeting, lend comfort and warmth to any living space.

Drywall and Paneling

You'll need do something about your ceiling, unless you're planning for the open loft. In the old days, the ceiling, like the floor, was built of rough sawn lumber, laid end to end, side by side. These boards, and rafters or beams supporting them, were often whitewashed to brighten up the room.

Drywall makes a pretty good alternative. Sold in 4x8 or 4x12 sheets, its goes up relatively fast. Once up, you can paint it or cover it up with paneling.

If you plan on dividing up your cabin space into smaller rooms, drywall and/or paneling offer quick solution for wall construction, too.

The one problem about drywall, though is in the finishing work. After you've nailed those sheets in place, you have to hide the seams, with drywall tape and layer upon layer of joint compound. Working with joint compound is tougher and more mundane than doing the chinking: you have to apply this white muck in layers, and sand each layer before you apply the next. It can seem more trouble than it's worth.

If you must sand, try wet sanding with a damp sponge. It's not nearly as messy as dry sanding, which can raise a fine white dust you'll never really clean up.

Of course you could forget the sanding and just nail paneling or trim over the drywall, or put up textured paint or acoustic tiles to hide the seams on the ceiling.

Porches and Decks

The front porch was a standard feature on any traditional log cabin. It typically claimed the entire side of the cabin surrounding the main entranceway, and was little more than an extension of the roof, supported by posts. Some had wooden plank floors, many did not.

Today, porches can be just as simple, or much more ambitious. A porch or deck remains one of the most enjoyable features of any home in the country, and every cabin should have one, or both, or some combination of them.

When planning, make them sufficiently wide to allow room for tables, chairs and an unimpeded walkway. Screening in the porch, or a section of it, will allow you to enjoy being outside without the annoyance of mosquitoes, moths and other summer bugs. A room-sized deck, added to the back of the house, gives you more usable space, and some privacy.

When you build your porch or deck, be sure to use treated lumber for the framing and decking. Porches and decks are constantly exposed to the weather and the long-term durability of this wood is well worth any additional cost.

Roofing

Unless your cabin needs to be historically accurate, don't bother with a traditional shake roof. If you must, frame and build a modern roof, than add your shakes.

As far as the roofing goes on a cabin, you can put down a shingle roof or a tin roof. A shingle roof is sold in squares, enough roofing to cover a 10x10 area. You'll need 4x8 sheets of sheathing for the entire roof, and then a layer of felt, before you can tack down the shingles. In about 20 years you can expect to need to replace those shingles.

I prefer a tin or "galvanized" roof for a several reasons: it goes down fast, it lasts longer than a shingle roof and I thoroughly enjoy the sound of rain falling on a tin roof. In the long run, its cheaper and much easier to maintain. If you have a choice of painted or unpainted tin, spend the little bit extra for the painted galvanized tin.

Tin roofing come in sheets wide enough to cover two or three feet at a time. It's also sold in standard lengths, from 6 to 16 feet, but depending on where you buy it you may be able to cut it to any length.

If you decide on a tin roof, use the roofing screws sold with rubber washers to hold it down. I've tried every type of roofing nail, and they just don't hold the tin down, or hold up as well as the screws.

Skylights

A skylight, properly installed, can transform the dark of the typical cabin into a bright, cheery space. I think every home can use at least one. There's a certain charm to sitting in your home at night and looking up and seeing the moon through the clouds or stars over head.

Skylights come in a variety of shapes and sizes, made of all types of glass. They are easily installed, but just as easily a source of leaks in the roof. If you are going to get one, you'll be better off building it into the roof, than cutting into the roof later. Caulk it thoroughly, watch for leaks, and always have plenty more caulk or roofing tar available.

Insulation

A tight log cabin, with all the gaps in the chinking plugged can be surprisingly well insulated against the weather outside, once the room warms up or cools down.

Today, that's not enough, though. You'll save a lot of energy by installing insulation above your ceiling or in the attic, and between the rafters of your floor. Depending on where you live, the local power company should be able to provide recommendations on how much insulation you need.

Fiberglass insulation is sold in rolls and pads. The rolls are sold by width to fit between studs or rafters set 16 or 24 inches apart. You can also buy the rolls in precut lengths, for the standard eight foot wall. For the attic or ceiling you're better off buying the insulation in an uncut roll. That way you can cut it to length, as needed.

I also recommend that you buy your insulation with paper backing, as it's much easier to handle and install. Even so, this stuff can be irritating: wear a dust mask, safety glasses gloves and a long sleeve shirt and long pants when you handle it.

One more bit of advice: when installing insulation under the cabin, be sure and protect it so your lovable dog or cat can't get to it. I spent a cramped week pushing through the crawl space under my house insulating all the floors, only to find a month or two later our dogs had been pulling a section down each night to make themselves comfortable new beds.

Next time, I stapled chicken wire over the rafters to protect the insulation and help hold it in place.

Appendix C

Other Methods for An Authentic Looking Log Cabin

There's more ways to create a cabin than the four methods I used. Some way want to consider these other options before you embark on your project. Or, perhaps you could combine these with some of the methods described earlier for a truly unique home.

Log Home Kits

Several companies help people fulfill their dreams of a log cabin by selling them a log home kit. These kits range from pre-hewn and notched logs for a cabin shell to all you need for the complete home, including recommendations on crews who can put it together.

Most of these companies specialize in a particular region of the country, offering both the type of woods and styles prevalent there. Look and ask around and you shouldn't have much trouble finding one or more of these kit companies in your area.

Try to compare what's offered from at least two vendors. Before you sign any contract, thoroughly assess all you get for your money: what the package does and doesn't include, the condition of the logs when they arrive, how they will be delivered, what the plans for the kit cover, any warranty on the treated logs, and possible expenses not covered by the kit price.

Also, do your own calculations on what the kit will really cost you. How much will you need to invest, in time, money and materials to transform that kit into a livable cabin?

BUYING OLD LOGS

If you live where log cabins were once a fixture on the landscape, you may be able to find and use logs from cabins which have been dismantled. They may not be easy to locate, but if you ask around and advertise you should come across a source.

Prices can vary, depending on the length of the logs and size of original cabin. Expect to pay at least $25 per log, and likely more. If the logs haven't been tagged before the structure was taken down, forget about trying to rebuild the original cabin.

More likely, though, you'll be buying logs which originally came from several different cabins. Get logs of ample length, as you'll need to cut off the old notches before you can make yours. You'll need to combine two or more logs to come up with full-length logs if you want a larger cabin or log home.

THE SAW MILL

There's a saw mill operating in most rural, wooded areas of the country. If you live near one, you can speed the process along, and save a considerable amount of work by buying your logs directly from a saw mill operator or logger.

If you plan on building something like the rounded log cabin from Chapter 3, estimate how many and what size logs you need. Then contract with a logger to have them delivered to your site.

If you want to build something which looks like a hand hewn cabin, have them cut at the saw mill. Use your own logs, or purchase them through the saw mill.

Simply have the opposite sides of the log run through the saw to give you a log of desired thickness. After the logs are squared off this way, you can "rough up" the sides with an axe to give them a hand-hewn look.

After that, it's just a matter of cutting your notches and completing all the other work that goes into a cabin.

THE TIMBER FRAME CABIN

Another time-saving method to create a cabin: build with precut timbers.

I've seen different cabins of this type built using 4x6, 6x6 and 6x8 beams. There's not much to chink, as the "logs" lay flat on top of each other. You simply caulk or chink the gaps between them to keep out drafts.

As far as the corners go, you can build with or without notching the ends. Some simply butt the end of one beam against the other, then spike them together. Or, let one end of alternate beams run past the end of the corner in each course, overhanging it 4 to 6 inches for the appearance of a notch .

If you prefer a real notch, use a simple lap joint or saddle notch. Both logs should extend slightly past the corner for more effect.

THE FAUX CABIN

I don't know what else to call it, but I've seen a number of creative solutions for capturing the look of a log cabin, without all the work. These "faux cabins" are often used for restaurants and souvenir shops in log cabin country, to catch the attention of passing tourists. It works!

One method involves framing the building using standard procedures, then covering it with plywood or wafer board sheathing. Once that's done, nail standard 2x6 or 2x8s to the plywood, evenly spaced to create the appearance of gaps between the logs. Occasionally chinking or caulk is applied to this space, but usually it's just painted white. Done right, from a distance it looks like a cabin.

You can make it even more realistic by substituting slabs of wood for the standard 2x6s or 2x8s. For this you would have a saw mill cut two inch thick slabs from full length logs, which you can nail to the building and rough up for a hand-hewn look.

For a rounded cabin, substitute the first slabs the mill cuts from logs when preparing to cut them up into usable lumber. If you talk to the saw mill operator you may be able to buy these slabs for little or nothing, as they are usually burned or sold by the load for kindling.

Resources

These are some resources I've turned to again and again, and are a good starting point for expanding your knowledge or finding the tools for your log cabin project.

The Foxfire Book
by Elliot Wiggington, Anchor Books
The original volume in the series devoted to the traditional lifestyle and crafts of the Appalachian region, including excellent chapters on building a log cabin and a shake roof.

How To Build and Furnish A Log Cabin
by W. Ben Hunt, John Wiley & Sons
The classic guide to building and furnishing a log cabin.

Basic Construction Techniques for Houses and Small Buildings Simply Explained
Prepared by the Bureau of Naval Personnel, Dover Publications
Best all-in-one resource of its type I've come across.

Basic Carpentry Techniques, Basic Wiring Techniques, Basic Plumbing Techniques
Ortho Books
Series of well illustrated, easy to understand guides on the basic skills and methods

Lehman's Non-Electric Catalog
Excellent source for many old-fashioned, hard-to-find items including some tools
(888) 438–5346. www.lehmans.com

Cumberland General Store
Another source of old time items you may have thought you would never see again, but available here, including some tools.
800–334–4640. **www.cumberlandgeneral.com**

Perma-Chink Supply, Inc.
Supplier of chinking solutions, sealants, finishes, preservatives and renovation tools and supplies for the log cabin.
www.perma-chink.com. Check website for phone numbers of regional distributors

Ebay: If you can't find what you need anywhere else you can probably find it on the online auction house. **www.ebay.com**

Questions, Comments? Email us @ **atticcorner@dtccom.net**

0-595-24571-4

Printed in the United States
1045900003B